Iris Murdoch Today

Series Editors
Miles Leeson
Iris Murdoch Research Centre
University of Chichester
Chichester, UK

Frances White
Iris Murdoch Research Centre
University of Chichester
Chichester, UK

The aim of this series is to publish the best scholarly work in Murdoch studies by bringing together those working at the forefront of the field. Authors and editors of volumes in the series are internationally-recognised scholars in philosophy, literature, theology, and related humanities and interdisciplinary subjects. Including both monographs and contributed volumes, the series is scholarly rigorous and opens up new ways of reading Murdoch, and new ways to read the work of others with Murdoch in mind. The series is designed to appeal not only to Murdoch experts, but also to scholars with a more general interest in the subjects under discussion.

Lucy Oulton

Iris Murdoch's Wild Imagination

Nature and the Environment

Lucy Oulton
Iris Murdoch Research Centre
University of Chichester
Chichester, West Sussex, UK

ISSN 2731-331X   ISSN 2731-3328   (electronic)
Iris Murdoch Today
ISBN 978-3-031-87832-9   ISBN 978-3-031-87833-6   (eBook)
https://doi.org/10.1007/978-3-031-87833-6

© The Editor(s) (if applicable) and The Author(s), under exclusive license to Springer Nature Switzerland AG 2025

This work is subject to copyright. All rights are solely and exclusively licensed by the Publisher, whether the whole or part of the material is concerned, specifically the rights of translation, reprinting, reuse of illustrations, recitation, broadcasting, reproduction on microfilms or in any other physical way, and transmission or information storage and retrieval, electronic adaptation, computer software, or by similar or dissimilar methodology now known or hereafter developed.
The use of general descriptive names, registered names, trademarks, service marks, etc. in this publication does not imply, even in the absence of a specific statement, that such names are exempt from the relevant protective laws and regulations and therefore free for general use.
The publisher, the authors and the editors are safe to assume that the advice and information in this book are believed to be true and accurate at the date of publication. Neither the publisher nor the authors or the editors give a warranty, expressed or implied, with respect to the material contained herein or for any errors or omissions that may have been made. The publisher remains neutral with regard to jurisdictional claims in published maps and institutional affiliations.

Cover illustration: © Alex Cooper / Alamy

This Palgrave Macmillan imprint is published by the registered company Springer Nature Switzerland AG.
The registered company address is: Gewerbestrasse 11, 6330 Cham, Switzerland

If disposing of this product, please recycle the paper.

For Jim

Acknowledgements

There are a great many people to whom I owe an enormous debt of gratitude for their generous support of this book which began as a PhD thesis. I would like to thank Miles Leeson for his unwavering belief, practical suggestions and calm advice and Hugh Dunkerley for his guidance in helping to secure the ecocritical underpinning to this book and my approach to the poetry. I am grateful to Paul Hullah who, together with Yozo Muroya, first brought Iris Murdoch's published poetry to critical attention. Thank you, too, to Tasha Alden for her insightful observations and enthusiasm in the lively viva discussion of this work.

More than a decade has gone by since I first studied Murdoch with Anne Rowe at Kingston University; I feel very fortunate to be able to continue to benefit from Anne's expertise, wisdom and acute observation. I want to thank the team at the Iris Murdoch Research Centre, University of Chichester, for their support and invaluable suggestions, in particular, Frances White for her knowledge and expertise, shared on many walks in the Surrey countryside, which have helped to shape my approach to Murdoch. At the Iris Murdoch Collections at Kingston University Archives and Special Collections, I owe archivist Dayna Miller an immense debt of gratitude for her knowledge, patience and steadfast support of this research. I wish to thank all of the transcribers who give so generously of their time to help make the materials more accessible to researchers all over the world, in particular, Rachel Hirschler for her commitment and valuable insights.

To become a Murdoch scholar is to become part of a warm and supportive academic family: this is an immense privilege. I am grateful for the suggestions, astute criticisms and recommendations offered to me at Murdoch conferences, postgraduate events and via email that have benefitted my work, and for the opportunity to publish my work in *The Murdochian Mind* and *Études britanniques contemporaines*—particular thanks to Hannah Marije Altorf, Lucy Bolton, Gary Browning, Silvia Caprioglio Panizza, Gillian Dooley, Camille Fort, Rob Hardy, Mark Hopwood, Marie Laniel, Megan Laverty, Rebecca Moden and Daniel Read. The online Iris Murdoch Book Club was formed during the Covid-19 pandemic and, thanks to technology (more on that later), participants from many different countries studied all twenty-six of Murdoch's novels over a two-year period together. Sincere thanks to Maria Peacock and all the Murdoch enthusiasts who participated in this initiative. This book benefits from those discussions.

It has been exciting to take Iris Murdoch among the ecocriticism community and to have travelled to Orkney and to Plymouth for the Association for the Study of Literature and the Environment Conferences in 2018 and 2019. Online events, including Adeline Johns-Putra's webinar 'Realism(s) in the Anthropocene: Representational and Ethical Challenges', organized by the European association; an ecocriticism course led by Jenny Bavidge at the Institute of Continuing Education, Cambridge University; and a workshop, 'Ethical Attention: Iris Murdoch in philosophical dialogue', a joint initiative of the Centre for Ethics in Public Life, University College Dublin, and the Centre for Ethics, Univerzita Pardubice, have all played an important part in bringing this project to fruition.

Last but not least, I would like to thank Jim, Sam, Katie, Mae, Emily, Michael, Jack and my canine research assistant Tess (2008–2022) for their love: each of them has helped me look at the world in new and different ways.

Competing Interests The author has no competing interests to declare that are relevant to the content of this manuscript.

Praise for *Iris Murdoch's Wild Imagination*

"This highly original monograph extends the well-established critical field of work on Murdoch's view of attention as the ultimate moral value to a new area, ecocriticism. Drawing on a remarkable breadth of material, Oulton outlines the development of Murdoch's thinking about the natural world in her fiction, philosophy and poetry across her whole career, in a ground-breaking study which opens up new avenues for Murdoch studies and literary ecocriticism."
—Tasha Alden, Senior Lecturer in Contemporary British Fiction, *Aberystwyth University, UK*

"In the current resurgence of interest in Iris Murdoch, many sides of her thinking have been reviewed, but this is the first book on her sense of nature and the environment. It is an outstanding work, perceptive in considering the multiple ways in which Murdoch thinks about nature. It shows how Murdoch explores the power, order and mystery of nature in her poetry, novels and prose writing. A 'must read' for Murdoch enthusiasts, and all who value a critical appreciation of nature and the environment."
—Gary Browning, Emeritus Professor of Political Thought, *Oxford Brookes University, UK*

"Lucy Oulton's groundbreaking study transcends that vital requirement that academic work make a substantial original contribution to knowledge in its field. Her sensitive and impressively eclectic (its subject would have approved!), agile-minded ecocritical reading of Murdoch is compelling, completely convincing, and full of meaningful surprises: magical moments of freshest insight abound. Murdoch is the gift that keeps giving—seek, and you will find it—but Oulton purposefully parses Murdoch's oeuvre with keen and reverent methodical critical incision. There is love in every line."
—Paul Hullah, Professor of British Literature, Meiji Gakuin University, Japan; poet, co-editor of the authorized *Poems by Iris Murdoch* (1997), and President of *The Iris Murdoch Society of Japan*

Contents

1. Iris Murdoch's Wild Imagination: Ecocritical Approaches — 1

2. Iris Murdoch, Poet: Ecological Themes and Lyrical Influences — 43

3. The Emerging Animism in the Novels: *The Flight from the Enchanter*, *An Unofficial Rose*, *The Unicorn* and *Nuns and Soldiers* — 95

4. The Sovereignty of the Sea in *The Sea, The Sea* — 147

5. Maurice Merleau-Ponty, Embodied Mind and Vegetal Agency: *The Good Apprentice* — 183

6. A Vision of the World as Sacred: Further Thoughts on Murdoch's Ecological Consciousness — 225

Index — 241

About the Author

Lucy Oulton holds a PhD and is a research associate at the Iris Murdoch Research Centre, University of Chichester. Her publications include 'Nature and the Environment' in Silvia Caprioglio Panizza and Mark Hopwood (eds.) (2022) *The Murdochian Mind*, and 'Loving by Instinct: Environmental Ethics in Iris Murdoch's *The Sovereignty of Good* and *Nuns and Soldiers*' in *Études britanniques contemporaines* 59, 2020. She edits the *Iris Murdoch Review* and volunteers at the Iris Murdoch Collections, Kingston University Archives and Special Collections.

Abbreviations

Items from the Iris Murdoch Collections at Kingston University Archives and Special Collections

Journals

Abbr.	Item	Dates	Archive no.
J3	Journal 3	4 June 1945–12 May 1947	KUAS202/1/3
J7	Journal 7	January 1949–1 January 1953	KUAS202/1/7
J9	Journal 9	30 March 1954–24 February 1964	KUAS202/1/9
J10	Journal 10	February 1964–18 March 1970	KUAS202/1/10
J11	Journal 11	18 March 1970–16 May 1972	KUAS202/1/11
J12	Journal 12	1 April 1975–23 May 1978	KUAS202/1/12
J14	Journal 14	1 January 1981–8 August 1992	KUAS202/1/14

Short-form notes from Murdoch's journals referenced in this book have been restored to full text for clarity and this is indicated in the quotations by square brackets.

Poetry Notebooks

Abbr.	Item	Dates	Archive no.
PN4	Poetry Notebook 4	22 January 1938–July 1940	KUAS202/3/4
PN5	Poetry Notebook 5	22 July 1940–June 1945	KUAS202/3/5
PN6	Poetry Notebook 6	1952–1962	KUAS202/3/6
PN7	Poetry Notebook 7	1972–1980	KUAS202/3/7
PN8	Poetry Notebook 8	23 May 1972–April 1974	KUAS202/3/8
PN10	Poetry Notebook 10	1972–1977	KUAS202/3/10

Murdoch's Libraries

Books from Murdoch's Libraries at the Iris Murdoch Collections are identified in the text by IML (Oxford Library) or MLL (London Library).

Works by Iris Murdoch

Novels

BP	The Black Prince	1973
FFE	The Flight from the Enchanter	1956
FHD	A Fairly Honourable Defeat	1970
GA	The Good Apprentice	1985
GK	The Green Knight	1993
HC	Henry and Cato	1976
MP	The Message to the Planet	1989
NG	The Nice and the Good	1968
NS	Nuns and Soldiers	1980
PP	The Philosopher's Pupil	1983
SPLM	The Sacred and Profane Love Machine	1974
TS	The Sandcastle	1957
TSTS	The Sea, The Sea	1978
TU	The Unicorn	1963
UN	Under the Net	1954
UR	An Unofficial Rose	1962
WC	A Word Child	1975

Philosophical Writings

SRR	*Sartre: Romantic Rationalist* (1967 and 1999 editions)
EM	*Existentialists and Mystics: Writings on Philosophy and Literature*
MGM	*Metaphysics as a Guide to Morals*

Essays from *EM* are referenced in the text as follows:

AD	Against Dryness
EH	The Existentialist Hero
EM	Existentialists and Mystics
FS	The Fire and the Sun: Why Plato Banished the Artists
IP	The Idea of Perfection
LP	Literature and Philosophy: A Conversation with Bryan Magee
NM	The Novelist as Metaphysician
OGG	On 'God' and 'Good'
SBR	The Sublime and the Beautiful Revisited
SGC	The Sovereignty of Good over other Concepts
TL	Thinking and Language

Previously published poetry collections

PIM	Muroya, Yozo, and Paul Hullah, eds. 1997. *Poems by Iris Murdoch.* Okayama: University Education Press
YB	Murdoch, Iris, and Reynolds Stone, *A Year of Birds*

Iris Murdoch Society Publications

The *Iris Murdoch Review* is available from the Iris Murdoch Society at the University of Chichester. Back copies are made available online.

*IMN*13	*Iris Murdoch Newsletter* 13	KUAS30/13	1999
*IMR*5	*Iris Murdoch Review* 5		2014
*IMR*10	*Iris Murdoch Review* 10		2019
*IMR*12	*Iris Murdoch Review* 12		2021

Other Works

Cue	Murdoch, Iris. 1970. A Note on Drama *Cue: Greenwich Theatre Magazine*, September 1970, 13–14, KUAS139/2, from the Iris Murdoch Collections
Giffords	Murdoch, Iris. 1982. Gifford Lectures: 2nd draft, manuscript KUAS202/6, from the Iris Murdoch Collections
GMH	Gardner, W.H., ed. 1953. *Gerard Manley Hopkins: A Selection of his Poems and Prose*. London: Penguin
IMAL	Conradi, Peter, J. 2001. *Iris Murdoch: A Life*. London: HarperCollins.
LoP	Horner, Avril, and Anne Rowe, eds. 2015. *Living on Paper: Letters from Iris Murdoch 1934–1995*. London: Chatto & Windus
MA	Mac Cumhaill, Clare, and Rachael Wiseman. 2022. *Metaphysical Animals: How Four Women Brought Philosophy Back to Life*. London: Chatto & Windus
MM-P	Maurice Merleau-Ponty, works by
OED	Stevenson, Angus. 2010. *Oxford Dictionary of English*, 3rd edn. Oxford: Oxford University Press
OLD	Lewis, Charlton T. and Charles Short. eds. 1879. *A Latin Dictionary*. Oxford: Oxford University Press
Podcast	a recording from the *Iris Murdoch Podcast* series available on Spotify, Apple and SoundCloud, see Miles Leeson et al.
RIP	Royal Institute of Philosophy lecture series
S&A	Conradi, Peter J. 2001. *The Saint and the Artist*. London: HarperCollins
TCHF	Dooley, Gillian, ed. 2003. *From a Tiny Corner in the House of Fiction: Conversations with Iris Murdoch*. Columbia SC: University of South Carolina

1

Iris Murdoch's Wild Imagination: Ecocritical Approaches

Introduction

Iris Murdoch's *Metaphysics as a Guide to Morals* (*MGM*) was published in 1992, a challenging and disparate text that was the culmination of almost a decade of excruciatingly hard graft to bring her 1982 Gifford Lectures to print and to a broader audience.[1] Although complex and often confounding, a remarkable feature of this philosophical work is its literary sensibility. Murdoch produces strikingly visual images whose power lingers long in the memory. One such picture emerges as she discusses the fact–value dichotomy. Values, she argues, have been sidelined by the dominance and widespread acceptance of perceived facts, stemming from the (albeit notable) advancements in twentieth-century science. Yet facts cannot be devoid of values; the one cannot operate independently of the other. Murdoch insists that 'values pervade and *colour* what we take to be

[1] For an illuminating biographical assessment of the genesis of the lectures and their development to publication, see Frances White (2019, 17–31); Megan Laverty and Evgenia Mylonaki are producing a new and 'resolute' reading, asserting the 'stunning ambition' of *MGM*. Far from being the 'rambling' work that some have suggested, *MGM* is organized, Laverty and Mylonaki argue, to be read not only by practising philosophers but by 'everyday metaphysicians' (Laverty 2023); for more on this topic, see Miles Leeson, Megan Laverty and Evgenia Mylonaki (*Podcast* 2023).

© The Author(s), under exclusive license to Springer Nature Switzerland AG 2025
L. Oulton, *Iris Murdoch's Wild Imagination*, Iris Murdoch Today,
https://doi.org/10.1007/978-3-031-87833-6_1

the reality of our world' (*MGM* 26). In this connection, she invites her reader to think of an image of Earth as seen from space:

> Think how our idea of our home planet has altered, both as we look back over hundreds of years, and over scores of years; Earth, now, as a travelling spaceship, seen from the outside, vulnerable, lonely, precious. (*MGM* 47)

Murdoch seeks to illustrate the degree to which humankind's perception of the meaning of existence has been altered by advancements in technology which, in turn, have contributed (and in her view not altogether favourably) to the modern demythologized age.

The first (dark, grainy, black and white) full-disc view of planet Earth was photographed from space on 30 May 1966 by the Soviet *Molniya 1–3* communications satellite (Uri 2020). Two years later, the famous 'Earthrise' photograph was shot by American astronaut William Anders; Anders would later be moved to remark on the ecological impact his image had had on him (Carlson 2024). Meanwhile back on Earth, poet Archibald MacLeish had only ever seen images broadcast in black and white, like everyone else in 1968, and simply marvelled at the astronauts' vantage point. 'To see the Earth as it truly is, small and blue and beautiful in that eternal silence where it floats', he enthused (O'Neil 2018). When Murdoch made her invocation in the 1980s, Earth-from-space images, although much improved in picture quality, were yet to be as readily ubiquitous as they are today. It is, of course, ironic that Murdoch's whole-Earth 'travelling space ship' (*MGM* 47) is only relatable to an audience that has been able to see such images on their television screen, aided by the sort of technology she sought to decry.[2] The magisterial beauty of the whole of Earth seen from space, once impossible, then exclusively the preserve of space travellers, now has the capacity to render us ordinary people awestruck too.

Murdoch embeds her fleeting visual metaphor into a complex philosophical argument that builds her case for a model of moralized visual

---

[2] In Murdoch's view, television 'can commit terrible crimes against the visible world'. She says, 'I am inclined to think that it blunts our general sense of colour and light and reduces rather than enhances our ability to see the detail of our surroundings' (*MGM* 330).

attention; there is an assumed shared value here in a deep reverence for Earth's precious beauty, inasmuch as she conveys her own anxiety about planetary isolation and vulnerability. Reinforcing the connection Murdoch makes, Cora Diamond suggests that,

> if we consider the taking in of the visual world with a kind of wonder and freshness of perception, a visual attention which can simply marvel at a shade of blue or at the twistedness of a tree trunk, which can take in the goodness and beauty of the world, then we do indeed have a model of moral awareness of reality. (1996, 108)[3]

For Murdoch, in the contemporary godless age, Earth viewed from space offers observers a renewed sense of the ineffable. Illustrating her text with planetary imagery in this way reveals, in macrocosm, a deep-rooted ecological consciousness of which we have only become more aware since Murdoch's personal journals and poetry notebooks joined the Iris Murdoch Collections at Kingston University Archives and Special Collections.[4]

Each one of Murdoch's remarkable journals serves as a 'capacious holdall'.[5] Each volume is richly populated with quotidian remarks about the seasons, favourite trees and observations of species and their behaviours littered among all manner of lengthy ruminations on philosophy and literature, notes on domestic life with John Bayley, interactions with friends and lovers, social occasions and travel. Within these pages, Murdoch records the passing of the seasons: 'The sun goes on and on shining [...]. The garden faintly surprised, summers away into deep autumn, the leaves fall slowly and meditatively' (J10 2). She observes the birds from her window, noting on 27 December 1968 that the 'garden is

---

[3] For a more recent assessment of Murdoch's 'Fact and Value' essay, see Craig Taylor (2019, 67–78).
[4] In September 2016, Audi Bayley donated Murdoch's personal journals and poetry notebooks to the Iris Murdoch Collections at Kingston University Archives and Special Collections; the volumes are located under KUAS202/1 and KUAS202/3, respectively.
[5] Thus Virginia Woolf expressed the purpose of her journal on 20 April 1919; I think it aptly describes how Murdoch made entries in hers even if, as we know, she did subsequently review and remove some content. Woolf notes, 'I should like it to resemble some deep old desk, or capacious hold-all, in which one flings a mass of odds and ends without looking them through' (1954, 13–14).

full of field fares [*sic*]' (J10 149). And she remarks on the affective experience of walking among the imposing trees in ancient woodland near her home in Oxfordshire (J14 111). Collectively, her observations reveal a life-long and profound engagement with the natural world, a preoccupation of the philosopher, novelist, poet and public intellectual that until recently has gone largely unexamined.[6] The expressions of ecological consciousness found in these fresh primary sources have inspired this new investigation of a range of her novels, her poetry and her philosophy.

Iris Murdoch's Wild Imagination opens up a rich new strand of enquiry for Murdoch studies by examining her work through an environmental lens, taking for its broader interrogatory framework the contemporary Earth-centred approach to literary criticism, known as ecocriticism. Ecocriticism, a relatively young interdisciplinary practice, interrogates representations of nature amid the immediate contemporary and overarching concerns about destructive human practices, the climate crisis, and Earth's capacity to sustain biodiverse life. This book is the first of its kind to scrutinize Murdoch's work at this critical intersection.

In the context of recent critical activity, Murdoch's fictional settings demand revisiting for their interpretive potential in the face of broader environmental concerns. Through her fictional characters, Murdoch connects sensory experience through the body with moral perception and this connection gains in significance when description of place in the novels is presented not merely as value-neutral setting but as fully immersive, thus making it morally significant. *Iris Murdoch's Wild Imagination* identifies this as Murdoch's ethics of place and, in its examination of relevant ecocritical theory, explores the phenomenological intersection of this theory with Murdoch's poetry and her unique brand of realist novel.

---

[6] On 9 October 2019, Lucy Oulton and Frances White presented a seminar on 'Iris Murdoch's Environmental Imagination' to postgraduate students at Université de Picardie Jules Verne, Amiens, as part of a lecture series, *Lieux revisités*. This event and the Iris Murdoch Conference in Amiens the following day culminated in two papers on Murdoch and nature being published in a special edition of *Études britanniques contemporaines*, titled *Iris Murdoch and the Ethical Imagination: Legacies and Innovations*. Lucy Oulton (2020) 'Loving by Instinct: Environmental Ethics in Iris Murdoch's *The Sovereignty of Good* and *Nuns and Soldiers*' and Frances White (2020) 'Anti-Nausea: Iris Murdoch and the Natural Goodness of the Natural World'.

Revisiting Texts, Revisiting Murdoch

In investigating Murdoch's environmental imagination, this book participates in the recently established process of reassessing place in literary texts against the backdrop of our very real contemporary concerns around climate emergency. 'Novels still have the power to affect the way societies see themselves', asserts Jenny Bavidge, as she observes the emergence of a period 'where a sense of ecological and climate crisis haunts stories which are not consciously environmental narratives at all' (Bavidge 2019, 1). To recast this idea, it would seem to constitute some form of oversight if the contemporary critic attempts an evaluation of place in a novel without considering its environmental or ecological space.

Such is the current anxiety regarding environmental degradation that conceptions of place and space and representations of the natural world in fiction seem to linger long in the minds of readers. In recent years, ecocritics have reassessed the work of some of our most preeminent British and Irish novelists, including Thomas Hardy, Virginia Woolf and James Joyce, and brought new perspectives to the poetry of William Wordsworth, John Clare, Gerard Manley Hopkins and W.B. Yeats, among others.[7] This book similarly reassesses Iris Murdoch to draw out the ecological consciousness evinced by her writings.

Murdoch may seem an unlikely choice for ecocritical evaluation when she is best known as a writer of urban novels that capture the mood of the post-war period and the second half of the twentieth century. The primary preoccupation of these novels is indisputably the human interaction of her generally middle-class, well-educated, suburban and city dwellers: she, herself, said that 'imaginative prose literature' is the 'form of art most concerned with the existence of other persons' (SBR 278).

---

[7] For example, Bonnie Kime Scott (2012) *In the Hollow of the Wave: Virginia Woolf and Modernist Uses of Nature*. Charlottesville, University of Virginia Press; Robert Brazeau and Derek Gladwin (eds.) (2014) *Eco-Joyce: The Environmental Imagination of James Joyce*. Togher, Cork University Press. Richard Kerridge (2001) 'Ecological Hardy', in Karla Armbruster and Kathleen R. Wallace (eds.) *Beyond Nature Writing: Expanding the Boundaries of Ecocriticism*. Charlottesville, University Press of Virginia, pp. 126–42; Jonathan Bate (1991) *Romantic Ecology: Wordsworth and the Environmental Tradition*. London, Routledge; John Parham (2010) *Green Man Hopkins: Poetry and the Victorian Ecological Imagination*. Amsterdam, Rodopi; Deborah Fleming (2019) '"All we know comes from you": W.B. Yeats and Ecocriticism', *Nordic Irish Studies* 18, 56–76 <https://www.jstor.org/stable/27041555> (accessed 29 November 2022).

Many of these novels are centred on her beloved London, including works such as *Under the Net* (1954), *The Time of the Angels* (1966), *The Black Prince* (1973), *A Word Child* (1975) and *The Green Knight* (1993). The novels that venture beyond London become noteworthy for their rich variety of alternative locations, for example, Paris in *Under the Net*, rural France in *Nuns and Soldiers* (1980), Ireland in *The Unicorn* (1963) and *The Red and the Green* (1965), England's unspecified northern coastline in *The Sea, The Sea* (1978), somewhere invented in *The Philosopher's Pupil* (1983), and somewhere both invented and quasi-mythic in *The Good Apprentice* (1985).

As novelist, Murdoch has garnered a significant reputation for depicting the complexities of some of the most human of preoccupations; freedom and servitude, fate and suffering, sexual freedom and gender identity, and how the moral life might function in a godless world represent some of her more common themes. By means of a unique form of psychological realism, her fiction is meticulously crafted to reveal the inner lives of her characters, often characterized by their complex social interactions, and these, in turn, invoke empathy in her reader. Fundamental to Murdoch's brand of realism, however, is a discernible ethics of place revealed in the powerful connection that exists between the inner lives of her characters and their lived environment. This is observable in key critical moments when (often solitary) characters interact with their environment.

In *Sacred Space, Beloved City: Iris Murdoch's London*, Cheryl Bove and Anne Rowe remark on Murdoch's ability to render London with an 'acuity which matches that of more celebrated "London writers", such as William Blake, Charles Dickens and Virginia Woolf' and the motivation that drives their book is to 'redress the balance and establish Murdoch amongst their ranks' (Bove and Rowe 2008, 1). Murdoch's foregrounding of the city seems almost celebratory, Bove and Rowe suggest, yet contrastive of her deep concern about the fragmentation of society wrought by two world wars, the terrible impact of human displacement and the concomitant heightened significance of place to the human psyche. They regard Murdoch's depictions of London as emblematic of her experimentation with form:

> descriptions of [London's] colours, sights, sounds, smells and movements provide an aesthetic moral apparatus which allows her readers to be

absorbed into the environment along with her characters, and allows them to participate equally in the sacred nature of the city. (Bove and Rowe 2008, 2)

In the London novels, Murdoch uses landmarks to make innovative and intricate connections between the 'environment and the human mind', where 'the city has the power to speak to the soul—her own, her characters' and her readers' alike' (Bove and Rowe 2008, 1). Bove and Rowe see the physical space of London as central to Murdoch's moral philosophy, in 'her attempt to encourage in her readers an awareness of what lies outside themselves, of the existence of the "Other" that pierces the fantasy world and leads to moral awareness' (2008, 2). The centrepiece of this philosophy is her affirmation of the Good as a substitute, in the contemporary age, for a personal God. Murdoch defines God in her 1969 essay, 'On "God" and "Good"', as being or having represented '*a single perfect transcendent non-representable and necessarily real object of attention*' (OGG 344).[8] This sovereign principle, she believes, can be applied to her conception of the Good; how one might pursue the Good is fundamental to the development of her philosophy and her fiction.

Murdoch understands literature, or the wider category of art, as analogous to morals. More broadly, she declares, 'We cease to be in order to attend to the existence of something else, a natural object, a person in need' (OGG 348). The viewing of an artwork or the reading of a novel itself can provide such revelation, she avers. Indeed, events within a novel can put a finer point on the matter, whether a 'natural object' or a natural scene. Through her use of realism, Murdoch provokes her characters into interactions with external landmarks or landscapes to confront their inner turmoil, and this is true, too, of the novels that go beyond London or, indeed, do not visit London at all.

---

[8] Murdoch's essay 'On "God" and "Good"' was first published in *The Anatomy of Knowledge* (Grene 1969, 233–58), a volume that published papers from meetings of the Study Group on Foundations of Cultural Unity (later, the Study Group on the Unity of Knowledge) organized by Marjorie Grene, Michael Polanyi and Edward Pols. The group was established in an attempt to unify centres of resistance to contemporary scientism. See Grene's brief commentary on Murdoch's essay (1969, 24).

As I have suggested, Murdoch expressed a profound distrust of the twentieth-century creep of scientism at the expense of moral perception. Her particular target was the precision of scientific explication that fails, in her view, to account for the nuance and complexity of moral life. In the 'Idea of Perfection', she avers that

> Words are the most subtle symbols which we possess and our human fabric depends on them. The living and radical nature of language is something which we forget at our peril. It is totally misleading to speak, for instance, of 'two cultures', one literary-humane and the other scientific, as if these were of equal status. There is only one culture, of which science, so interesting and so dangerous is now an important part. But the most essential and fundamental aspect of culture is the study of literature, since this is an education in how to picture and understand human situations. We are men and we are moral agents before we are scientists, and the place of science in human life must be discussed in *words*. This is why it is and always will be more important to know about Shakespeare than to know about any scientist: and if there is a 'Shakespeare of science' his name is Aristotle. (IP 326–27)[9]

Aristotle she admired for his intellectual range; he was concerned with ethics, metaphysics and, importantly, the philosophy of science. 'The Idea of Perfection' originates from the Ballard Matthews Lecture that Murdoch presented to the University College of North Wales in 1962; the essay was subsequently published in *The Yale Review* in 1964 and republished as the first of three essays in her volume *The Sovereignty of Good* in 1970. American marine biologist and poet Rachel Carson had made perfectly clear in her 1952 National Book Award acceptance speech for *The Sea Around Us* (1951) that ecology was being deeply impacted by this apparent divorce of science from the arts. Carson's remarks foreshadow C.P. Snow's influential lecture 'Two Cultures' that he delivered at Senate House, Cambridge, on 7 May 1959 (which itself expanded on Snow's article for the *New Statesman* on 6 October 1956). The citation

---

[9] For more on the significance of Murdoch's response to the exchange on 'Two Cultures' between C.P. Snow and F.R. Leavis, see Patricia Waugh, 'Iris Murdoch and the Two Cultures: Science, Philosophy and the Novel' (2012, 33–58).

for Carson's award described *The Sea Around Us* as a 'work of scientific accuracy presented with poetic imagination and such clarity of style and originality of approach' (Popova 2019, 436). Accepting her award, Carson said:

> The aim of science is to discover and illuminate the truth. And that, I take it, is the aim of literature, whether biography or history or fiction; it seems to me, then, that there can be no separate literature of science. […] The winds, the sea, and the moving tides are what they are if there is wonder and beauty and majesty in them, science will discover these qualities. If they are not there, science cannot create them. If there is poetry in my book about the sea, it is not because I deliberately put it there, but because no one could write truthfully about the sea and leave out the poetry. (Popova 2022)

Science without a proper account of ethics was of serious concern to Carson; for Murdoch it represented pure danger. Murdoch's writerly imagination foregrounds moral perception in her picture of human interactions with the external material world. At the same time, her mistrust of technology is revealed in fleeting allusions in the novels, whether they are directly related to environmental catastrophe, to characters discussing the gathering pace of technological progress or the mutually assured destructive potential of nuclear war, or to simply reveal her general resistance to the inevitability of the coming technological age.

References suggestive of environmental catastrophe begin early in the novels. In Murdoch's third novel, *The Sandcastle*, published in December 1957, schoolmaster William Mor is fixated by the nature of his infidelity: he has struck up a relationship with the beautiful young portrait artist Rain Carter, for whom he intends to leave his wife and children. He is waiting for Rain at Waterloo Station, ruminating obsessively over his desperate situation. He imagines that 'this is what it is to be one of the damned' (*TS* 234). He buys a newspaper: 'Dreadful headlines stared him in the face. *Bridegroom Kills Bride in Car Crash. Possible Contamination of Earth's Atmosphere: Scientists' Grave Warning*' (*TS* 234). The second of these headlines could conceivably allude to the disaster that struck the Sellafield nuclear power plant in the North West of England in the same year *The Sandcastle* was published, or certainly to the anxiety surrounding the establishment of such sites. Sellafield was the world's first commercial plant of its kind. Opened on 31 March 1956, it was heralded as a cheap source of electricity for the country while, in fact, being secretly tasked to produce plutonium for Britain's nuclear weapons programme. In October 1957, the Windscale Pile No. 1 erupted in flames. John Vidal of the *Guardian* recalls,

> it was hushed up so well that even with 11 tons of uranium ablaze for three days, the reactor close to collapse and radioactive material spreading across the Lake District, the people who worked there were expected to keep quiet and carry on making plutonium for the bomb. (Vidal 2012)

Britain's first nuclear power plant was cloaked in secrecy. Intriguingly, in *The Sandcastle*, as Mor reads these newspaper headlines, Murdoch's narrative embarks on a meditation on the nature of truth: on Mor's failure to be entirely honest with either his lover, Rain, or his wife, Nan. His personal crisis turns over and over relentlessly in his mind. He knows he must accept that 'he had now definitely and irrevocably parted company with the truth' (*TS* 235). Confronted with the bleak headlines and unable to bear them, Mor discards the newspaper.

Early on in *The Good Apprentice*, dinner party conversation at the McCaskervilles' house in Fulham comprises a considerable range of topics: nuclear war, disease, television and artificial intelligence. All of these are suggested, by some of those present, as potential threats to the human race (*GA* 29–36). Family doctor Ursula Brightwalton defends what she understands as the benefits of the new scientist age: 'thinking itself can be pure, without values, like genuine science, like maths' (*GA* 31). Harry Cuno commends the wisdom and efficiency of machines: 'Even now a machine can see infinitely more than we can, see it faster, discern more details, make more connections, correct itself, teach itself, learn new skills which we can't even conceive of', and he asserts that a 'machine is *objective*' (*GA* 30). Both characters, each in their own way, display a striking naïveté. Murdoch anticipates the increasingly complex issues that surround advancements in science and technology (notably, in artificial intelligence) and presents them as potential threats to the natural world, in the sense of what might stand to be lost through the indiscriminate backing of scientific progress. In her novels, Murdoch often includes characters who favour a plant-based diet, and she tends to use apparently perfunctory means to express contingent human impact, as is the case here, for example, in *The Sea, The Sea* with regard to discarded cans or broken bottles:

> Below the causeway, on either side, there is a wilderness of small rocks, piled higgledy-piggledy by nature, and not accessible to the sea. This is a less attractive scene and not without a few rusty tins and broken bottles which I must one day climb down and remove. (*TSTS* 11)

Murdoch's environmental concerns, while exhibited in plain view, are presented in an effortlessly incidental manner in her fiction.

Justin Broackes places anti-scientism in the top five of Murdoch's 'largest ideas for academic moral philosophy' in his introduction to *Iris Murdoch, Philosopher*, and says that Murdoch is motivated by the desire 'to escape the view that the world can be said to contain only what science tells us is there or what is clearly reducible to that' (2012, 8). In an essay in the same collection, Alison Denham declares that the 'beating heart' of Murdoch's philosophy is to be found in her thinking on moral perception—Murdoch's 'idea that moral experience features a quasi-experiential phenomenology analogous to sensory (and particularly visual) perception' (2012, 328). The reality of the world, as Patricia Waugh explains, includes 'values as part of the ordinary phenomenology of experience and these values might be understood as the very ground of our picture of what it is to be human: its deepest configuration' (2012, 45). Waugh argues that such values emerge from the interplay of the embodied self, bodily action and conscious experience. *Iris Murdoch's Wild Imagination* takes these phenomenological connections for its starting point in an interrogation of the embodied physical interactions that Murdoch creates between character and environment. In these moments, Murdoch reveals her conception of an animate world, the independent lives of other species and a profound belief in the affective impact of such a living, feeling environment on the human body and the mind.

As readers of Murdoch's novels, we witness direct interactions between character and environment where Murdoch accords full ethical importance to these transactional moments. These moments seem crucial to her neo-theological conception of the Good, when she asks this two-fold question: 'Can good itself be in any sense "an object of attention"? And how does this problem relate to "love of the real"?' (OGG 356). Murdoch is searching for an appropriate substitute for the practice of prayer:

> I am not here thinking of any quasi-religious meditative technique, but of something which belongs to the moral life of the ordinary person. The idea of contemplation is hard to understand and maintain in a world increasingly without sacraments and ritual and in which philosophy has (in many respects rightly) destroyed the old substantial conception of the self. A sacrament provides an external visible place for an internal invisible act of the spirit. Perhaps one needs too an analogy of the concept of the sacrament, though this must be treated with great caution. Behaviouristic ethics denies the importance, because it questions the identity of anything prior to or apart from action which decisively occurs, 'in the mind'. The apprehension of beauty, in art or in nature, often in fact seems to us like a temporally located spiritual experience which is a source of good energy. (OGG 356)

In this essay, Murdoch connects the spiritually uplifting quality of apprehending beauty with what is overlooked when ethics fails to take account of (or dismisses the significance of) 'anything prior' (OGG 356). Murdoch is acknowledging here the body's capacity to react to events in advance of the mind (Oulton 2014, 8). She recognizes there is much that needs to occur in the body physiologically before the mind decides to act.

The Murdochian account of the relationship between knowing and valuing, which at a practical level can be seen as lived experience, anticipates contemporary neuroscientific work. According to Patricia Waugh,

> Murdoch builds an argument that language arises out of the processes of the interactive embodied mind in the world and the way in which that complex processual system of gesture, sensation, emotion, perception, intuition, makes possible a more self-conscious linguistic representationalism, but one where language is always already 'soaked in' (TL 33) or 'riddled with the sensible' (TL 39), and is coextensive with, rather than the foundation of, thought. (2012, 43)

For Murdoch, sensation and language are co-dependent:

> Language itself, if we think of it as it occurs 'in' our thoughts, is hardly to be distinguished from imagery of a variety of kinds—hardly to be distinguished at times, one might add, from sensations, in the sense of obscure bodily feelings. (TL 39)

Embodied experiences are brought to bear, albeit often momentarily, on the ensuing moral imperative of her narratives; experiences are often expressed in a situated or embodied form of entanglement that appears to suggest that a sort of phenomenological exchange occurs. At critical moments in the novels, place is rendered epistemologically and morally significant. This is Murdoch's environmental imagination at work.

Connecting Murdoch's moral perception to sensory experience in this way helps explain why a text on Murdoch's cultural engagement with visual art is valuable to this discussion on Murdoch and the natural world. In *The Visual Arts and the Novels of Iris Murdoch*, Rowe makes the case for Murdoch's reliance on visual imagery in her novelistic art when words prove inadequate:

> Murdoch's cross-fertilization […] stems from a desire to use visual images as an extension of language, so that the innermost thoughts and feelings of her characters, inaccessible often to the characters themselves, can be *experientially* assimilated by readers through a purely instinctive sensual response, and not by means of sustained intellectual questioning. (Rowe 2002, 2)

This acute observation of Murdoch's sensory interaction with art can also be productively applied to the novelist's phenomenological engagement with the natural world and Earth's materiality. Murdoch delineates attending to art and attending to nature:

> There are many harmoniously unified forms in nature, but our pleasure here is not always a search for 'objects', but may be a delight in limitless continuation or accumulation or chaos: mountains, waterfalls, forests, the sea, the sky. Kant even saw this experience of 'the sublime' as having a superior spiritual function as contrasted with enjoyment of the 'beauty' of limited forms. (*MGM* 3–4)

Murdoch also acknowledges the challenge of attending to art when 'the mood of the new ecological and social utilitarianism can find the artificiality of art frivolous in a suffering world' (*MGM* 4). Clearly, Murdoch's novels (art) mediate nature. A brief moment in *The Green Knight* illustrates this mediation and is representative of scenes from a range of her novels that are constructed in a similar vein.

Moy, an aspiring artist disheartened by a deeply unsatisfactory meeting with her tutor Miss Fox, leaves the art school and walks 'at random' down to the 'eternal' River Thames (*GK* 174). The river is a distinct feature of many of Murdoch's London novels including *Under the Net*, *The Time of the Angels* and *A Word Child* and, according to Rowe, it represents the 'most powerful and ubiquitous London presence', its tidal fluctuations meticulously 'built into her novels' realism' (2019, 111). In the heart of London, at this preeminent landmark, Murdoch depicts a private moment of full-bodied interaction between Moy and the riverine environment:

> There was a fuzzy grey mist over the Thames, the tide was out, the narrow stream looked dull and sluggish, like thick grey oil oozing along. [...] She walked down over the sticky mud to the water's edge [...]. Here the sound of traffic had become a woodland murmur, remote from the calm pace of the eternal Thames. (*GK* 174)

Moy is brooding on the realization that she ought to have taken her 'wilder, more outrageous work' to impress the tutor (*GK* 174). Close to tears, she attempts 'to calm herself by standing very still and gazing at the fuzzy mist which was motionlessly pendant above the water' (*GK* 174). Suddenly, she is caught up in a commotion. Moy's attempt to intervene and rescue a small black duck under attack from a swan causes her to enter the freezing water. But the feathered aggressor turns its attention on Moy, 'pressing her down with its descending weight, as it had pressed down the little struggling duck', before it rises up over the water and flies away (*GK* 175). Distracted from the minutiae of the earlier frustration, that had led her to the river in the first place, the wet, muddy and extremely chilled Moy now sheds tears for fear that, in attempting to defend the duck, she may have injured the swan. The episode recalls Ovid's 'Leda and the Swan' but is also likely inspired by W.B. Yeats's poem of the same name. Yeats's poem presents a graphic retelling wherein the violated body of Leda is considered symbolic of the political situation in 1920s Ireland. Murdoch greatly admired Yeats and his influence on her was life-long. Published almost thirty years earlier, *The Red and the Green* (1965), for example, represents far more than 'vague homage to the great Irish poet', according to Gillian Dooley and Frances White, who observe the intertextuality of Murdoch's only historical novel with Yeats's 'Easter 1916', whereby Murdoch reworks the events of 1916 'in her own artistic image' (Dooley and White 2019, 1, 10–11). In *The Green Knight*, Murdoch's swan scene alludes to, rather than retells, the Leda story but its inclusion is nonetheless far from incidental. Murdoch reverses or, at the very least, rebalances the power dynamic of girl and swan. Like the Murdochian kestrel that I discuss shortly, the swan causes Moy's attention to divert from her own petty concerns to pay close attention to the events unfolding before her.

Moy's attempt to save the small black duck underscores the potential for opposing interpretations of her reaction—the sense of an attentive stewardship on the one hand, but an anthropomorphized response on the other—when she attempts to interpose herself between the independent lives of waterfowl ordinarily left to determine their own fate. This unpleasant intervention, fraught with danger, is encoded with two Murdochian conceptions at once: a realism that meticulously accounts for the other (as much the contingent messiness of the tidal river bed as the independent lives of creatures beyond her human characters) and, importantly, a moment for Moy of 'unselfing' that engenders attending to the other. For Murdoch, truthful vision is achieved by paying proper attention to the world outside oneself, and this scene is presented as just such a moment of truthful vision for Moy. Murdoch explains that the 'self, the place where we live, is a place of illusion. Goodness is connected with the attempt to see the unself' (SGC 376); but she also acknowledges the constant and hard work involved. Influenced by Simone Weil's concept of '*décreation*' (1952, 36), Murdoch aspires to these moments of unselfing in which one is required to attempt to forget the self in an effort to attend to the other. At the water's edge, Murdoch recreates a 'self-forgetful' encounter of girl and swan in an expression of unforced attention to natural beauty: 'the only spiritual thing we love by instinct' (SGC 369–70). The power of these moments lies in Murdoch's conception of a sensory world that induces self-reflection in her characters and, in turn, her reader. Murdoch's distinctive experimentation with the realism of place draws her reader into a vibrantly comprehensive picture, and it is such realist representations that engender ethical attention.

Ecocriticism and the Phenomenological Intersection with Murdoch

Iris Murdoch's Wild Imagination offers a unique contribution to the large and diversifying field of literary ecocriticism in its investigation of Murdoch and the natural world. A reflection, then, on the discipline's fast-moving emergence into the mainstream is essential in order to highlight the specific interests and practices that can be brought to bear on discussions at the critical intersection of ecocriticism and Murdoch studies. As a term, 'ecocriticism' is credited to William Rueckert who, with a paper titled 'Literature and Ecology: An Experiment in Ecocriticism', was set on making a 'contribution to human ecology, specifically, literary ecology' (1996, 105, 107), and did so in 1978. By the time Rueckert's paper was published in *The Ecocriticism Reader* in 1996, a fringe preoccupation within cultural studies in the United States was maturing into its own discrete field of enquiry. In the *Reader*'s introduction, co-editor Cheryll Glotfelty offers an early definition of the field: 'Ecocriticism takes as its subject the interconnections between nature and culture specifically the cultural artifacts of language and literature' (Glotfelty 1996, xix).[10] The *Reader* set out to engage at this intersection of ecological and literary practice.

In the late 1990s, Richard Kerridge defined ecocriticism for a British readership, importantly connecting the discipline to a developing awareness of environmental crisis:

> The ecocritic wants to track environmental ideas and representations wherever they appear, to see more clearly a debate which seems to be taking

---

[10] In 1996, the Association for the Study of Literature and the Environment (ASLE) was created in the United States; ASLE-UKI followed in the United Kingdom and Northern Ireland. The European Association for the Study of Culture, Literature and the Environment was established a few years later.

place, often part-concealed, in a great many cultural spaces. Most of all, ecocriticism seeks to evaluate texts and ideas in terms of their coherence and usefulness as responses to environmental crisis. (Kerridge 1998, 5)

Just over a decade later, Greg Garrard endorsed Kerridge's all-encompassing approach but took issue with what he saw as Kerridge's 'monolithic conception of "environmental crisis"' and his 'seemingly secure ecological yardstick' when ecology itself, 'both as a science and as a socio-political movement', is in flux (Garrard 2012, 4). Garrard's *Ecocriticism* includes a brief but stark assessment of the range of issues threatening the environment for ecocritics who 'may not be qualified to contribute to debates about problems in ecology, but [...] must nevertheless transgress disciplinary boundaries and develop their own "ecological literacy" as far as possible' (2012, 5). Kerridge's early efforts to help establish ecocriticism as a discipline in the United Kingdom, nonetheless, led him to outline the demands being made, by environmentalism, of literary criticism:

> Often, literature, especially narrative, is regarded as the cultural space reserved for the 'personal' viewpoint [...]. This notion of the personal tends to exclude the large-scale perspectives, political generalities, narrative time-scales and scientific vocabularies used in environmental debate. (Kerridge 1998, 6)

Nonetheless, he declares, ecocriticism should challenge literature to 'show how it feels, here and now' (Kerridge 1998, 6). Kerridge argues that literature has the capacity to dramatize the vital human sensory perception of the environment, and indeed its destruction—his 'here and now' (notably a quarter-century ago) emphasizing his own sense of the dire urgency in this regard (1998, 6).

For Garrard, ecocriticism's 'widest definition', effectively arising from the sense of one's self (the human) in relation to everything else (the environment), is the 'study of the relationship of the human and the non-human, throughout human cultural history and entailing critical analysis of the term "human" itself' (Garrard 2012, 5). Self-perception holds significant bearing on how one interprets the human relationship with the environment, and the number of identifiable environmental positions speaks to the knotty complexity of the issue. Garrard's *Ecocriticism* discusses many of them. For example, the 'cornucopian' view suggests that a plentiful Earth exists as resource for the benefit of humankind; ecofeminism shares many of the traits of the wider feminist movement and, as feminism resists the oppression of women, it resists the oppression of the natural world by humans driven by the human/male 'mastery model' (2012, 29); social ecology and ecomarxism apply the basic principles of their left-leaning ideologies to environmental matters; deep ecology values all living things equally. The difficulty of deep ecology, inherent in its aspiration towards equity for all living things, is the resulting apparent logical absence of responsibility or stewardship. Deep ecology places ecology, not humanity, at the centre of decision-making and is 'merely an "orientation"' for those that demand consideration for the 'intrinsic value' of nature, according to Garrard. However, this 'remarkable even-handedness might well seem to empty deep ecology of any substantive content: if value resides everywhere, it resides nowhere as it ceases to be a basis for making distinctions and decisions' (2012, 25).[11] Humanity has impacted Earth's environment, causing devastation in some quarters; humanity is in the unique position of being able to lead change in attitudes and practices. In practice, deep ecology in extremis would appear to represent a hopelessly rudderless and drifting worldview. What appear to be outright philosophical or political positions for some persist as orientations for others, adding a further layer of complexity and nuance to the environmental debate.

Ecocriticism has continued to evolve for more than a quarter-century in waves of theoretical development. Over time, the discipline has essentially become more accommodating of urban/rural discussions and nature/culture discourses, reaching back to Welsh literary critic Raymond Williams—*The Country and the City* being an early iteration of the contiguity and co-reliance of the urban and the rural—and forward to new conceptions around the definition of nature itself, and the additional complexity, for example, of debates around ethics and biopolitics. 'The country and the city are changing historical realities, both in themselves

---

[11] As a radical position, deep ecology appears controversially to adopt a 'misanthropic' view in its attempts to address matters such as human population (Garrard 2012, 25); for more on the range of eco-philosophical 'orientations' see Garrard (2012, 18–36).

and their interrelations', Raymond Williams famously wrote (2019, 415). Born into the same generation as Murdoch, it has proven difficult to establish whether Williams and she ever actually met; it could be said that their paths crossed on a couple of occasions, paths that were strictly speaking more political than environmental. In 1938, Murdoch won a League of Nations essay competition with the title 'If I were Foreign Secretary'; Williams came second (*IMAL* 78). They were both members of the Communist Party of Great Britain for a time, and two of the seventy signatories to the *May Day Manifesto 1968* which set out a new socialist agenda in the face of the failing post-war political consensus in the United Kingdom. What is clear is that both Murdoch and Williams backed the call for a new socialist Britain after the austerity of the war years.

In the United States, ecocriticism focused on non-fiction naturewriting at first, centring on wilderness and the 'intense individual connection with the landscape', an approach denounced by some as 'celebratory' (Marland 2013, 849, cites Head 2000, 236, and Barry 2009, 242) and, suggests Pippa Marland, one that displayed a 'relatively uncritical understanding of "nature"' (2013, 849). Unable to lay claim to any sort of wilderness per se in Britain, ecocriticism has had necessarily different beginnings. In Britain, the emphasis has been on championing works that focus on the non-human world that have the potential to 'foster environmental sensibility' and, as such, British Romantic poetry was brought under renewed scrutiny (Marland 2013, 849). To this end, in the 1990s, Jonathan Bate reclaimed Wordsworth, Clare and others in two significant works, *Romantic Ecology: Wordsworth and the Environmental Tradition* and *The Song of the Earth*. Marland suggests that Bate's approach foregrounds a sense of poetry's potential for 'being in the world that is receptive to the self-disclosure of nature' (2013, 849). She explains,

> Bate characterises *ecopoetry* as a *phenomenological* and pre-political form, which draws us into communion with the earth through its emphasis on 'presencing' rather than representation, bodying forth that presencing in part through its rhythms and sounds. (Marland 2013, 849)

Bate's 'pre-political' approach to poetry (if one can ever truly escape the political) seems to accord with Murdoch's own phenomenological engagement with the environment.[12] Poetry's capacity to mediate nature in this way resonates with Murdoch's conception of the Earth as sacred, and in the manner in which she gives expression to an immersive embodied presence within the environment in some of her early poetry and in her novels. It is important to be clear that 'sacred' in the Murdochian context of treasuring Earth (or, indeed, in connection to her beloved London) is used here in the secular sense to mean 'regarded with great respect and reverence' (*OED*).

But what exactly do ecocritics mean by nature? Second-wave ecocriticism began to look more critically at conceptions of nature, and the constructs of nature in culture (or how culture constructs nature), with publications such as Kate Soper's *What is Nature?* and Timothy Morton's *Ecology Without Nature*.[13] These nuanced approaches contribute to a wider sense of the complexity of nature not as a thing apart but where the human co-exists with the non-human in an entangled weave of life and sharing Earth's biosphere. Morton is the more forthright on the matter, arguing that it is our concept of nature, itself, that 'impedes a proper relationship with the earth and its life forms' (2007, 2). Placing 'something called Nature on a pedestal and admiring it from afar does for the environment what patriarchy does for the figure of Woman'; such an approach, they argue, is a 'paradoxical act of sadistic admiration' (Morton 2007, 5). Morton contends that the 'time should come when we ask of any text, "What does this say about the environment?"', and they propose an 'ambient poetics' that provides a 'materialist way of reading texts with a view to how they encode the literal space of their inscription' (2007, 5, 3). This position fuses writer and environment, collapsing subject–object duality. At the launch of their book, *Being Ecological*, at the Royal Society of Arts in London on 29 January 2018, Morton reminded us of the simple fact that 'we live in a social space that was never really

---

[12] By 'pre-political', I take Bate to mean 'not consciously political', but he appears to rely on the status quo or a normatively accepted position, which of course, in itself, is a political one.

[13] Lawrence Buell formally ascribes waves to the development of ecocriticism (2005, 21–22).

human in the first place'.[14] In appearing to suggest that nothing today is nature, they appear to argue that everything, in fact, is exactly that.

What nature constitutes precisely should not be allowed to obscure ecocriticism's governing motivation: to decentre the human. Ecologists and ecocritics have devised means of referring to those aspects of the world that are not human including the obvious term, non-human. Used ubiquitously, however, non-human starts to determine what another being is not in relation to the human and, in that sense, can constitute an (often unintended) expression of anthropocentrism or, worse, human primacy. Ecologist and philosopher David Abram's term 'more-than-human' (1996, 22) not only provides for the differing potentials of other creatures to possess qualities that are not available to humans, but also suggests a world with scope for independent existence beyond humanity. Abram advocates a reassessment of animate Earth, whose happenings are registered in the mind through the feeling body. He considers the sentience of other creatures and, inspired by the work of Maurice Merleau-Ponty, is a proponent of the worldview that sees a continuous reciprocity between 'our sensory perceptions' and the feeling Earth: 'an interpenetrating webwork of perceptions and sensations borne by countless other bodies' (Abram 1996, 65). For Abram, this amounts to 'the biosphere as it is experienced and *lived from within* by the intelligent body—by the attentive human animal who is entirely a part of the world that he, or she, experiences' (1996, 65). Abram's conception of a sensuous world chimes with Murdoch's own vision of humanity as constituents 'of a living, palpable planet, connected not only to its physical actuality and temporal motion, but also to its eternal ethereal mystery', as Rowe encapsulates it (2019, 112). As I have suggested elsewhere, 'sometimes tight into the margins of her fiction, [Murdoch] pictures an "age-old reciprocity with the many-voiced landscape" (Abrams 1996, 5) and conveys her own sensuous conception of a shared animate world' (Oulton 2020, §13). Phenomenological engagement of this sort is key to understanding Murdoch's environmental imagination.

---

[14] For a recording of the book launch of *Being Ecological*, <https://www.youtube.com/watch?v=d_5UWI-SEVE> (accessed 30 March 2024).

Environment, as a term, has to work hard in these sorts of discussions; it is arguably controversial owing to the inherent anthropocentrism in its meaning. Environment 'presupposes an image of man at the centre, *surrounded* by things' (Bate 2000, 107), and yet it is ubiquitous and liberally deployed in climate-related discussions today. As Bate's discomfort reveals here, the term stands to reinforce the anthropocentric picture, when a move towards an all-encompassing, ecocentric worldview ought to lie at the heart of any discussion. 'Ecocentrism' (also known as 'biocentrism') suggests a 'worldview that sees all of nature as having inherent value, and is centred on nature rather than on humans' (Allaby and Park 2013, 131). Ecocentrism's target is to decentre the human. Yet, environment also 'relates to the natural world and the impact of human activity on its condition' as well as aiming 'to promote the protection of the natural world' (*OED*). It is precisely through this kind of self-conscious re-evaluation that second-wave ecocriticism began to lend greater voice to environmental justice and to offer its own corrective to the seemingly firm-set dualities that Marland lays out for her reader: 'culture/nature, male/female, mind/body, civilized/ primitive, self/other, reason/matter, human/nature and so on' (2013, 852). As a result, a more immersive approach to ecocriticism has emerged at the interchange between body and environment, and a general examination of being in the world, of which Abram is a proponent. Here, too, is where Murdoch's phenomenological engagement belongs.

The Body in the World, the Body in the Realist Novel's 'Storyworld'

The ecocritical potential of the novel form started to gain traction during the second wave of ecocriticism. For some it was an essential move: 'If ecocriticism is to realise its full potential, it will need to find a way of appropriating novelistic form' (Head 2000, 236). According to Marland, the interest lay in the novel's 'more self-conscious textuality', positioned as it was to reveal the 'complex entanglement of self and world, social and environmental history' (2013, 851–2). As ecocritical practice has matured, the wave analogy ascribed to the discipline's evolution has foundered on the increasingly diverse range and simultaneity of ideas that have grown up particularly around materiality and post-humanism. These ideas question the whole notion of boundary setting or hierarchy—for example, between human and animal, or rock, or tree—in persistent efforts to decentre the human.

Material ecocriticism concerns the co-existent and enmeshed forms of the world that constantly re-form by force, agency or other matter. Louise Westling, like Abram before her, draws on the philosophy of Maurice Merleau-Ponty in her work on material ecocriticism, particularly to express 'human immersion in nature and the meaning immanent in ordinary experience' (Westling 2011, 126). Connected to these material considerations is a concept that Jane Bennett calls 'thing power' to describe the agentive power of the non-human in the complex co-dependent web of life (Bennett 2010, 2). The resultant hybridity graphically takes on new temporal and spatial implications in the 'slow violence' of, for example, plastic waste finding its way into the stomachs of sea birds, mammals and fish where 'man-made substances find new agential roles' (Marland 2013, 858). 'Slow violence', according to Rob Nixon, is defined as the sort of devastation that 'occurs gradually and out of sight, a violence of delayed destruction that is dispersed across time and space, an attritional violence that is typically not viewed as violence at all' (2011, 2). In emphasizing the complex web of interactions that comprises Earth's ecology of which the human forms merely a part, material ecocriticism's first concerns have been to divest people of the instinctive impulse to human primacy over other species and at the same time to raise awareness of the ineradicable impact of humans on Earth. Murdoch's animistic portrayal of landscape speaks to her own instinctive awareness of just such a web of interactions.

Affect theorists prioritize the role of the body in responding to events but until fairly recently have not attended particularly to the way in which environments produce embodied affective response. In a similar way, ecocritics initially paid scant attention to the affective possibilities of the body's response to environment. Like ecocriticism, affect theory comprises several orientations, and Kyle Bladow and Jennifer Ladino note the synergy in simultaneous turns to affect and to the material in ecocriticism (2018, 4). Both theoretical groups recognize the 'structure of feeling' discerned by Raymond Williams, just over half a century ago, as providing the germinal moment for this intersection of affect theory and the material strand of ecocriticism (Williams 2019, 39, 49, 64). 'All matter […] is "storied matter"', according to Serenella Iovino and Serpil Oppermann who characterize 'the world's material phenomena' as 'knots in a vast network of agencies, which can be "read" and interpreted as forming narratives, stories' (Iovino and Oppermann 2014, 1). Like Bennett, Iovino and Oppermann ascribe an agential force to the more-than-human—not only to animals and birds, but trees, rocks, stones and so forth. Affect theorists investigate the encounter with, and the response to, that force: a force that is connected to love. For instance, Alexa Weik von Mossner discusses Martha Nussbaum's definition of love: 'the intense attachment to things outside the control of our will' (Nussbaum 2015, 15, cited in Weik von Mossner 2018, 51), and Nussbaum's belief—one we can recognize as aligning with Murdoch's own view—that such feelings have the potential to produce empathy in the fight against injustice and exploitation.[15] Working at the intersection of ecocriticism and affect theory, Weik von Mossner argues that our environment is fundamental to the way we feel and, reciprocally, fundamental to the way we feel about the environment. There is, then, a clear connection here between Murdoch's ethics of place and affect theory's concerns with the reciprocity of feeling and emotion between humans and nature.

Murdoch's environmental imagination helps to shape the distinctive brand of realism that, in turn, has the power to move her reader, with whom there is implicit understanding of a boundless world unconstrained by interpretative possibilities beyond the pages of her narratives. As Adeline Johns-Putra explains, 'a realist novel's "storyworld" must potentially and hypothetically include the entire world, in order to retain its status as realism' (Johns-Putra 2019, 9, cites Oak Taylor 2019, 35). Erin James says that 'even those [texts] that do not seem to be interested in the environment in and of itself, offer up virtual environments for

---

[15] Murdoch scholars will recognize Nussbaum's alignment to Murdoch's conceptions of love and attention, using Murdoch's 'M and D' example in making her case (Nussbaum 2015, 15); see also IP 312.

their readers to model mentally and inhabit emotionally' (2015, 54). Jesse Oak Taylor asserts that any 'work we read now is situated within the Anthropocene, regardless of where or when it was written' (2019, 40). His particular interest lies in 'literary atmosphere' which takes account of the conditions of the literal as well as the literary landscape in the development of character and plot. Hans Ulrich Gumbrecht deploys *Stimmung*—the German term that manages to encapsulate both mood and atmosphere:

> Reading for *Stimmung* [...] means paying attention to the textual dimension of the forms that envelop us and our bodies as a physical reality—something that can catalyze inner feelings without matters of representation necessarily being involved. (Gumbrecht 2012, 5)

How the body feels the literal reality or, by extension, the literary reality of environment has secured a place for affective ecology in the ecocritical landscape.

The first quarter of this century has also witnessed global discussion of an important proposal that suggests that Earth has left behind the relatively stable conditions of the Holocene—marking a period of relative stability in the Earth's climate dating from around 11,000 years ago following the last major ice age—and entered a new geological interval: the Anthropocene (the age of humans).[16] In 2000, the Anthropocene was proposed as the stratigraphic term to acknowledge the impact of human activity on the Earth.[17] But, if Earth has entered a new geological epoch, when did this actually begin? Candidates for human events impacting Earth's increasingly unstable conditions included the Agrarian Revolution (or more accurately, the profusion of agricultural developments over tens of centuries); the nineteenth century's Industrial Revolution; and the mid-twentieth century's Great Acceleration, characterized by the bulk

---

[16] See *British Geological Survey* (2023) 'Impacts of Climate Change'.

[17] Stratigraphy is the branch of geological studies concerned with strata (rock layers) and stratification (layering); the stratigraphic term 'Anthropocene' was proposed in 2000; see P.J. Crutzen and E.F. Stoermer (2000) 'The Anthropocene', 1–20.

rise in carbon dioxide levels, private vehicle use, intensive farming and energy consumption, among other markers of ostensible progress.[18] Timothy Clark explains why all this is important:

> To recognize the Anthropocene as 'emergent' alters the understanding of what may be environmentally destructive or not. For the encroachment of human activity on more and more of the biosphere is often a result of activities that once straightforwardly enhanced human welfare but which have now crossed a certain threshold in magnitude and impact. (2015, 48)

The International Union of Geological Sciences (IUGS) deliberated on this stratigraphic proposal for fifteen years. Eventually, on 20 March 2024, IUGS declared that for all the reasons outlined in their findings, the Anthropocene would remain an 'invaluable descriptor in human-environment interactions' but did not in fact mark a new geological time unit.[19] The 'invaluable descriptor' seems, nonetheless, here to stay; environmental scientists and ecocritics continue to use the term to confront the implications of humanity's impact on Earth and to convey the urgency of the climate crisis.

Researchers are mining centuries of georgic writing for insights into the most fundamental of human preoccupations: 'How should we work to cultivate a fertile and sustainable relationship with our physical and social environment? How, in other words, are we supposed to live well?' (Erchinger et al. 2021, 1). At the same time, ecocritics such as Timothy Clark and Derek Woods have attempted to contextualize the vast temporal and spatial scales that overwhelm literature because of the innate tendency to configure most fictional works to the human lifespan.[20] Clark argues that modes of interpretation must evolve to confront the pervasive and intractable issues brought about by the Anthropocene—he calls this 'scale framing' (Clark 2015, 73). He says,

---

[18] See J. Zalasiewicz et al. (2015) 'When did the Anthropocene begin?'
[19] For the full statement, see *International Union of Geological Sciences* (2024) 'The Anthropocene'.
[20] See Derek Woods 'Scale Critique for the Anthropocene' (2014, 133–42), and Timothy Clark (2015, 71–80).

Scale effects in particular defy sensuous representation or any plot confined, say to human-to-human dramas and intentions, demanding new, innovative modes of writing that have yet convincingly to emerge. (Clark 2015, 80)[21]

In his book *The Great Derangement: Climate Change and the Unthinkable*, Amitav Ghosh points to a remarkable concurrence of the evolution of literature with the build-up of 'carbon in the atmosphere [...] rewriting the destiny of the earth', remarkable, because of the distinct absence of climate discourse (Ghosh 2016, 7). Literary art, in Ghosh's view, is severely compromised by this absence. Johns-Putra expresses the temporal conundrum: 'scientific questions of climate occur at a very different spatio-temporal scale from the individual concerns of the literary' (2019, 7). At the same time, she acknowledges that 'narrative literary form is not incapable of imagining time above and beyond the human', reminding us of Frank Kermode's analogy of the 'day to day (or *chronos*) and the cosmological (*kairos*)' but, here, Johns-Putra draws a distinction between 'cosmological' and 'geological' time (2019, 9–10). In her view, 'kairotic time goes hand in hand with theological conceptions of the human' while the geological time reflected by issues relating to climate change 'demands a reckoning with non-human destiny, a reckoning that the still ubiquitous forms of literature, such as fiction and poetry, are only now beginning to perform' (Johns-Putra 2019, 10). Johns-Putra's work examines the capacity of realism to address these diverse spatio-temporal demands while it embeds emotional experiences of its characters into richly drawn settings. This study seeks to demonstrate how Murdoch makes her own

---

[21] Some will argue that the novel form, owing to its human-centric nature, cannot address the vast complexity of what is at stake; however, an emerging genre of novel often referred to as climate fiction (cli-fi) attempts to confront issues of scale hitherto overlooked. For example, the temporal setting of Richard Powers's novel *The Overstory* opening in mid-nineteenth century in America and spanning several generations is actually measured in 'tree years' (or concentric rings), a scalar dimension that by comparison miniaturizes the human lives depicted. Late in the novel we are told a couple have, for years, comforted themselves by reading through '*The Hundred Greatest Novels of All Time*' on which the narrator comments, 'To be human is to confuse a satisfying story with a meaningful one, and to mistake life for something huge with two legs. No: life is mobilized on a vastly larger scale, and the world is failing precisely because no novel can make the contest for the *world* seem as compelling as the struggles between a few lost people' (2018, 382–83).

temporal demands of her readers by contrasting spatial and temporal scales of landscape with the quotidian demands of character. Drawn together, these represent key elements of Murdoch's own unique brand of psychological realism.

It is important to emphasize that *Iris Murdoch's Wild Imagination* is not attempting to suggest that Murdoch's fiction overtly dramatizes environmental threats to individual lives, nor does this book claim that she particularly foregrounds the natural world in her work, expressly writing the environment into her novels. What Murdoch does do, as demonstrated in the works chosen for discussion in the following chapters, is to immerse her characters in a fully represented, contingent, animistic world, and she expresses how this world feels. At key moments she writes the sensory effects of wild places into her characters' lives, creating a phenomenological conception of their situatedness. She depicts crucially important epiphanic moments for her characters, who appear to experience (sometimes only momentary) unselfing alone in nature. Sophie Grace Chappell describes epiphany as arriving 'not by argument or deduction, but by our directly or immediately seeing or otherwise experiencing it. [...] [T]hey are moments of vividly experienced present-ness or attentive awareness, in which, as we sometimes say, time seems to stand still' (Chappell 2022, 4), and this description encapsulates nicely those moments in Murdoch novels that I regard as epiphanic. In the novels, these occasions manifest as brief intuitive opportunities to grasp reality. At the same time, Murdoch accounts for the independent lives of species outside the human sphere. Overall, there is something discernible in Murdoch's approach to the natural world and green spaces more generally that appears reverential. She takes account of our human dependence on the natural world, both physically and psychologically, which reveals a poet's awe for Earth itself. Murdoch tends to add spatial and temporal scales to the often relatively trivial quotidian demands of her characters that together serve as a sobering reminder to her reader of human insignificance in the context of a vastly older planet.

'The Universal Omnivorous Writer'

Any assessment of Murdoch's ecological consciousness calls for evaluation across the range of her literary, philosophical and personal writing to explore some of the influences on her depictions of place and her environmental thinking. This is the first major study to appraise a range of Murdoch's poems—a few have been previously published, others are published here for the first time—which focus on the natural world and on our human relationship with it, an urgent ethical issue in the twenty-first century. The purpose is to identify links that help establish this early poetic craft as an important prequel to the story of her subsequent novel-writing. This book also considers a range of her fiction, some of her philosophical writing and excerpts from her unpublished personal journals that offer insights into her environmental imagination. By including the range of her work in this way, *Iris Murdoch's Wild Imagination* makes its own contribution to an ongoing and important debate in Murdoch studies.

Some have argued that Murdoch the philosopher and Murdoch the novelist are separate enterprises, while others do not see these parallel endeavours so neatly categorized or corralled, especially when Murdoch herself seemed equivocal on the issue.[22] When Murdoch asserts in interview that philosophy and literary art are 'quite different operations', it certainly provides evidence to support the notion that her fiction-writing and philosophy are discrete activities (Sagare 2001, 697). Here, she suggests that 'it's very dangerous if a novelist attempts to express a philosophy or definite theory in a novel' (Sagare 2001, 697). Yet, her position appears inconsistent when she tells Stephen Glover, 'I'm not a philosophical novelist, and the philosophy really comes in rather incidentally' (*TCHF* 42). To Bryan Magee she reveals 'an absolute horror of putting theories or "philosophical ideas"' into her fiction:

---

[22] For the most recent assessment of the complexities of this long-held debate on Murdoch as philosopher-novelist, see Miles Leeson (2022, 363–75); for an important contemporaneous discussion of Murdoch's own views on the two disciplines of literature and philosophy see LP 3–30.

> I might put in things about philosophy because I happen to know about philosophy. If I knew about sailing ships I would put in sailing ships; and in a way, as a novelist, I would rather know about sailing ships than about philosophy. (LP 19–20)

The answer she offers Jack Biles, this time about coal mining, is similarly enigmatic: 'I mention philosophy sometimes in the novels because I happen to know about it, just as another writer might talk about coal mining; it happens to come in' (*TCHF* 58). Yet, even this pronouncement ends with a definitive statement: 'No, I wouldn't say I'm a philosophical novelist', and there the whole debate turns full circle (*TCHF* 58). Given the apparent ambiguity of Murdoch's own position it is unsurprising that some Murdoch scholars elect to take a more fluid approach in appraising her oeuvre where enquiry moves between her philosophy and fiction, the two categories representing a dialogic enterprise.

Daniel Read is aware of the potential controversy in interpreting Murdoch's writings in this way but adds, I think, helpful clarity to the debate when examined from the perspective of his own work on the problem of evil:

> the argument that [Murdoch] portrays a dialectical picture of morality [in her fiction] may appear to conflict with her absolutist moral philosophy. If her picture of the moral life is centred around a Platonic Form of the Good, an objector might ask, surely evil represents a failure in goodness and thus represents a force that cannot be engaged with in the moral life? In turning towards Murdoch's aesthetics, however, the reader finds a secure place for such moral thinking: art provides an arena 'where everything under the sun can be examined and considered' (FS 461) and should, therefore, necessarily reflect upon the realities of the moral life, including cruelty, suffering and wickedness. (Read 2019, 4)[23]

Gary Browning is convinced that Murdoch's work belongs together as an integrated whole, the one genre of writing reliant upon the other. He argues that the

---

[23] Daniel Read's monograph *Degrees of Evil in Iris Murdoch's Fiction and Philosophy* is to be published by Palgrave Macmillan in 2025.

literary exploration of the modern world matters to philosophy because the lived experience of the modern world is the object of philosophical analysis. Murdoch's metaphysics analyses the meaning of lived experience and her philosophical reading of modern forms of experience goes hand in hand with the imagined modern worlds of her novels. The one understands and theorizes what the other discloses. (2018, 55–56)

Indeed, in *The Murdochian Mind* on the subject of ecology, I argue that 'Murdoch's fictional acuity helps to reveal her environmental ethics to her reader' (Oulton 2022, 453). Despite Murdoch's resistance to the idea that there could be any sort of interrelationship between her own fiction and philosophy, she nonetheless appears to have regarded the writing of philosophical novels as a suitable endeavour. In her revised introduction to *Sartre: Romantic Rationalist* (1953) published in 1987, Murdoch expresses her admiration for the existentialist philosopher for just this sort of engagement: 'Sartre is, in himself, as philosopher, novelist, playwright, literary critic, biographer, essayist, journalist, a remarkable instance of the universal omnivorous writer' (*SRR* 1999, 12). Malcolm Bradbury thinks Murdoch's book on Sartre is 'striking' emphatically because, 'although it is much concerned with Sartre's version of post-Hegelian philosophy, and his views on ethics, metaphysics and politics, she comes to consider these matters largely through the route of his novels' (2000, ix). For Bradbury, *Sartre: Romantic Rationalist* is 'a meditation on the nature of the novel, its close relation to philosophy, and what the novel can instinctively do' (2000, ix). Even before she began publishing novels, Murdoch was mulling over the dialogical potential of philosophy and art. In her original introduction to *Sartre* published in 1953, Murdoch observes that:

> The novelist proper is, in his way, a sort of phenomenologist. He has always implicitly understood, what the philosopher has grasped less clearly, that human reason is not a single unitary gadget the nature of which could be discovered once for all. The novelist has his eye fixed on what we do, and not on what we ought to do or must be presumed to do. (*SRR* 1967, 8)

Sartre, she continues, 'brings to the novel, together with a remarkable literary gift, his typically philosophical self-consciousness'; Murdoch brings to the novel, alongside her own remarkable gift, a typically philosophical self-awareness. Her approach to novel-writing was experimental. Bradbury avers that, beginning with *Under the Net*, 'Murdoch would write philosophical novels challenging in one way or another the idea of the philosophical novel' (2000, xi). Indeed, on the subject of her first novel, in *Metaphysics as a Guide to Morals*, Murdoch steps back across her own barricade, ironically in a (customary) dismissal of Jacques Derrida, when she refers her readers to *Under the Net* for a 'philosophical discussion of these matters' (*MGM* 187). Where philosophers and critics can perhaps agree is that Murdoch understood the potential of the novel as a vehicle for playing out philosophical conundrums. Miles Leeson's thoughtful assessment of the current debate nonetheless advises a level of caution in moving across genres: 'Much is at stake in reading Murdoch's work in both forms, and we must be careful not to conflate ideas, or to transpose ideas between genres, however tempting it may be' (Leeson 2022, 368). The study of Murdoch now comprises more and diverse sources: her personal journals, her letters and, more recently, her poetry notebooks, so more care than ever is required.[24]

Any objection to an all-encompassing approach to Murdoch's oeuvre cannot overlook the one singular and formidable intellect at the source of this large body of published and unpublished writing. With the Iris Murdoch Collections' acquisition of poetry notebooks, personal journals and asymmetric letter-runs in recent years, it seems inevitable that Murdoch studies should evolve to take account of some of the more biographical elements of her writing. The publication of a number of biographies on Murdoch has inevitably presaged a determined move in this direction, as has the publication of private letters assembled into a volume with biographical and critical commentary.[25] A comprehensive but careful reading of her journals, letters, novels, philosophy and poetry as one writer's output can only help illuminate Murdoch's ecological consciousness.

---

[24] Leeson (2022, 374n) acknowledges White's essay as the 'best work highlighting the discussion between the fiction and philosophy in [Murdoch's] journals'; see Frances White (2019, 17–31).

[25] For example, Peter J. Conradi (2001) *Iris Murdoch: A Life*. London, HarperCollins; A.N. Wilson (2003) *Iris Murdoch: As I Knew Her*. London, Arrow; David Morgan (2010) *With Love and Rage: A Friendship with Iris Murdoch*. Kingston-upon-Thames, Kingston University Press; and Frances White (2014) *Becoming Iris Murdoch*. Kingston-upon-Thames, Kingston University Press. Avril Horner and Anne Rowe (2015) *Living on Paper: Letters from Iris Murdoch 1934–1995*. London, Chatto & Windus.

About This Book

Blending the biographical with the lyrical, *Iris Murdoch's Wild Imagination* explores how Murdoch's early poetry directs us to the first iterations of her ecological consciousness, a theme that has to compete with a multitude of others later on in the novels. *Iris Murdoch's Wild Imagination* charts Murdoch's early development as a reader and writer of poems at school and during her time at Oxford. I discuss a few of her poems, some published here for the first time, in the context of existing Murdochian poetry criticism and current ideas in environmental criticism and poetry on nature. The chapter also explores some of the poets that so obviously inspired the young Murdoch. I seek to make no particular claim about Murdoch as a poet. Instead, I focus on the evidence of Murdoch's attention to and love of the natural world that emerges early on in her literary practice through the more personal medium of poetry and carries through into her fiction.

Murdoch's early poetry presages the close attention she pays to nature's smallest details in the early novels. Her evident sense of the equity of all life is explored and, indeed, is confirmed by the emerging animism of the later fiction. Close attention to such small details gives rise to a profound sense of belonging for one character in *The Flight from the Enchanter* (1956). In this novel and in *An Unofficial Rose* (1962) a pattern of solitary moments in nature begins to emerge that becomes a common feature in all of the novels discussed in this book. Aesthetic considerations of the formal garden in *An Unofficial Rose* give way to primordial landscape in *The Unicorn* (1963). As Murdoch's novels become more expansive, so begins the accommodation of vaster scales of space and time that counterbalance the early minute details. The agentive qualities of the lithic landscape of the Provençal mountains in *Nuns and Soldiers* (1980) are discussed in this context. Murdoch's conception of the equity of all life and her animistic sensibility are connected to love and represent a hitherto unexplored constituent of a broader Murdochian ethics and, as such, has the potential to contribute to broader contemporary ecological discussions.

Iris Murdoch's Wild Imagination reassesses the sea and its eponymous material presence as the central component or hero of Murdoch's novel in a chapter on *The Sea, The Sea*. It counters critical analysis of the novel that casts the sea as empty and the sea's chief purpose to provide a psychological metaphor for the narrator's inner life and chaotic relationships. Instead, I consider the presence—and the evidence of implied presence— of rich and abundant sealife; the constancy of the sea's tidal waters; and the manner in which Murdoch asserts the sea's material presence in resistance to her narrator's solipsistic and sovereign demands on those (both human and non-human) around him. With reference to the Romantic movement, as well as the blue ecology theory that forms part of contemporary ecocriticism, this book offers a comprehensive reading of the novel that foregrounds the vital littoral protagonist—the sea—taking the phenomenology of ocean for what it is, and beyond its presence in the novel as the amorphous by-product of its narrator's turmoil.

Iris Murdoch's Wild Imagination then returns to the land to focus more intently on Murdoch's depictions of key moments of human interaction in landscape just as I have exemplified with Moy, the swan and the silty Thames shoreline at low tide in *The Green Knight*. Now, with reference to *The Good Apprentice*, I begin by charting Murdoch's early fascination with the role of the body and sensory awareness as essential precursors to thought. I consider Murdoch's early interest in the work of Maurice Merleau-Ponty and examine the degree to which his ontology of 'wild being' might be applied in any practical or meaningful sense. I assess interactions between character and animistic landscape in the novel in order to scrutinize the collision of agential forces that seems to occur at these powerful moments. Murdoch's portrayals of such interactions are suggestive of the affective experience of body in landscape that is recognized as critically significant by today's affect theorists. An associated theoretical discourse, adopted by practitioners in the field of material ecocriticism, constitutes vegetal agency which asserts the agential force of nature. Such notions of vegetal agency inform discussions of how Murdoch often pictures spaces neglected or abandoned by human inhabitants that take on an overwhelming sense of rewilding, constituting an agentive force: the vegetal on the creep.

In its entirety, *Iris Murdoch's Wild Imagination* comprises a new direction for Murdoch studies. It aims to provide a significant entry point for a broader discussion of Murdoch and the natural world and to demonstrate her relevance to environmental criticism. The research for this book has revealed the potential for related areas of ongoing scrutiny and I conclude with some thoughts about ongoing and further research.

References

Primary Sources: Novels

Murdoch, Iris. 1994. *The Green Knight* (*GK*) (1993). London: Penguin.
———. 1999. *The Sea, The Sea* (*TSTS*) (1978). London: Vintage.
———. 2000. *The Flight from the Enchanter* (*FFE*) (1956). London: Vintage.
———. 2000. *An Unofficial Rose* (*UR*) (1962). London: Vintage.
———. 2000. *The Unicorn* (*TU*) (1963). London: Vintage.
———. 2000. *The Good Apprentice* (*GA*) (1985). London: Vintage.
———. 2001. *Nuns and Soldiers* (*NS*) (1980). London: Vintage.
———. 2003. *The Sandcastle* (*TS*) (1957). London: Vintage.

Primary Sources: Philosophical Writings

Murdoch, Iris. 1967. *Sartre Romantic Rationalist* (SRR 1967) (1953). London: Fontana.
———. 1997. 'The Fire and the Sun: Why Plato Banished the Artists' (FS). In Iris Murdoch, *Existentialists and Mystics: Writings on Philosophy and Literature*, ed. Peter J. Conradi, 386–463. London: Penguin.
———. 1997. 'The Idea of Perfection' (IP). In Iris Murdoch, *Existentialists and Mystics: Writings on Philosophy and Literature*, ed. Peter J. Conradi, 299–336. London: Penguin.
———. 1997. 'Literature and Philosophy: A Conversation with Bryan Magee' (LP). In Iris Murdoch, *Existentialists and Mystics: Writings on Philosophy and Literature*, ed. Peter J. Conradi, 3–30. London: Penguin.
———. 1997. 'On "God" and "Good"', (OGG). In Iris Murdoch, *Existentialists and Mystics: Writings on Philosophy and Literature*, ed. Peter J. Conradi, 337–362. London: Penguin.

———. 1997. 'The Sublime and the Beautiful Revisited' (SBR). In *Iris Murdoch, Existentialists and Mystics: Writings on Philosophy and Literature*, ed. Peter J. Conradi, 261–286. London: Penguin.
———. 1997. 'The Sovereignty of Good over other Concepts' (SGC). In *Iris Murdoch, Existentialists and Mystics: Writings on Philosophy and Literature*, ed. Peter J. Conradi, 363–385. London: Penguin.
———. 1997. 'Thinking and Language (TL). In *Iris Murdoch, Existentialists and Mystics: Writings on Philosophy and Literature*, ed. Peter J. Conradi, 33–42. London: Penguin.
———. 1999. *Sartre Romantic Rationalist* (SRR 1999) (1987). London: Vintage.
———. 2003. *Metaphysics as a Guide to Morals (MGM)*. London: Vintage.

Primary Sources: Journals from the Iris Murdoch Collections

Murdoch, Iris. February 1964–18 March 1970. Journal 10 (J10). KUAS202/1/10.
———. 1 January 1981–8 August 1992. Journal 14 (J14). KUAS202/1/14.

Secondary Sources: Blogposts, Books, Chapters, Journal Articles, Newspaper Articles, Podcasts, Websites

Abram, David. 1996. *The Spell of the Sensuous: Perception and Language in a More-Than-Human World*. New York: Vintage.
Allaby, Michael, and Chris Park, eds. 2013. *A Dictionary of Environment and Conservation*. 2nd ed. Oxford: Oxford University Press.
Barry, Peter. 2009. *Beginning Theory: An Introduction to Literary and Cultural Theory*. 3rd ed. Manchester: Manchester University Press.
Bate, Jonathan. 2000. *The Song of the Earth*. London: Picador.
Bavidge, Jenny. 2019. A Sense of Climate Crisis Now Haunts Stories Which Aren't Even About the Environment. *The Conversation*, September 24. https://theconversation.com/a-sense-of-climate-crisis-now-haunts-stories-which-arent-even-about-the-environment-123429. Accessed 24 September 2019.
Bennett, Jane. 2010. *Vibrant Matter*. Durham NC: Duke University Press.
Bladow, Kyle, and Jennifer Ladino, eds. 2018. *Affective Ecocriticism: Emotion, Embodiment, Environment*. Lincoln NE: University of Nebraska Press.

Bove, Cheryl, and Anne Rowe. 2008. *Sacred Space, Beloved City: Iris Murdoch's London*. Newcastle upon Tyne: Cambridge Scholars Publishing.

Bradbury, Malcolm. 2000. Introduction. In Iris Murdoch, *The Philosopher's Pupil* (1983). London: Vintage.

British Geological Survey. 2023. Impacts of Climate Change. https://www.bgs.ac.uk/discovering-geology/climate-change/impacts-of-climate-change/. Accessed 26 April 2023.

Broackes, Justin, ed. 2012. *Iris Murdoch, Philosopher: A Collection of Essays*. Oxford: Oxford University Press.

Browning, Gary. 2018. *Why Iris Murdoch Matters*. London: Bloomsbury Academic.

Buell, Lawrence. 2005. *The Future of Environmental Criticism: Environmental Crisis and Literary Imagination*. Oxford, Blackwell Publishing.

Carlson, Michael. 2024. William Anders Obituary. *Guardian*, June 9. https://www.theguardian.com/science/2024/jun/09/william-anders-obituary. Accessed 18 July 2024.

Chappell, Sophie Grace. 2022. *Epiphanies: An Ethics of Experience*. Oxford: Oxford University Press.

Clark, Timothy. 2015. *Ecocriticism on the Edge: The Anthropocene as a Threshold Concept*. London: Bloomsbury.

Conradi, Peter J. 2001. *Iris Murdoch: A Life (IMAL)*. London: HarperCollins.

Crutzen, P. J., and E. F. Stoermer 2000. The Anthropocene. *IGBP Global Change Newsletter* 41. http://www.igbp.net/news/opinion/opinion/haveweenteredtheanthropocene.5.d8b4c3c12bf3be638a8000578.html. Accessed 14 March 2018.

Denham, A. E. 2012. Psychopathy, Empathy, and Moral Motivation. In *Iris Murdoch, Philosopher*, ed. Justin Broackes, 325–352. Oxford: Oxford University Press.

Diamond, Cora. 1996. "We Are Perpetually Moralists": Iris Murdoch, Fact, and Value. In *Iris Murdoch and the Search for Human Goodness*, ed. Maria Antonaccio and William Schweiker, 79–109. Chicago: University of Chicago Press.

Dooley, Gillian, ed. 2003. *From a Tiny Corner in the House of Fiction: Conversations with Iris Murdoch (TCHF)*. Columbia SC: University of South Carolina.

Dooley, Gillian, and Frances White. 2019. 'A Terrible Beauty': Iris Murdoch's Irish Novel. *The Red and The Green*', English Studies. https://doi.org/10.1080/0013838X.2019.1672449.

Erchinger, Philipp, Sue Edney, and Pippa Marland. 2021. Eco-Georgic: From Antiquity to the Anthropocene. An Introduction. *Ecozon@* 12 (2): 1–17. https://doi.org/10.37536/ecozona.2021.12.2.4537.

Garrard, Greg. 2012. *Ecocriticism*. 2nd ed. Abingdon: Routledge.

Ghosh, Amitav. 2016. *The Great Derangement: Climate Change and the Unthinkable*. Chicago: University of Chicago Press.

Glotfelty, Cheryll. 1996. Introduction: Literary Studies in an Age of Environmental Crisis. In *The Ecocriticism Reader: Landmarks in Literary Ecology*, ed. Cheryll Glotfelty and Harold Fromm, xi–xxxvii. Athens GA, University of Georgia Press.

Grene, Marjorie, ed. 1969. *The Anatomy of Knowledge*. London: Routledge & Kegan Paul.

Gumbrecht, Hans Ulrich. 2012. *Atmosphere, Mood, Stimmung: On a Hidden Potential of Literature*. Translated by Erik Butler. Stanford CA: Stanford University Press.

Hämäläinen, Nora, and Gillian Dooley, eds. 2019. *Reading Iris Murdoch's Metaphysics as a Guide to Morals*. London: Palgrave Macmillan.

Head, Dominic. 2000. Ecocriticism and the Novel. In *The Green Studies Reader: from Romanticism to Ecocriticism*, ed. Laurence Coupe, 235–241. Abingdon: Routledge.

International Union of Geological Sciences. 2024. The Anthropocene, March 20. https://www.iugs.org/_files/ugd/f1fc07_40d1a7ed58de458c9f8f24de5e739663.pdf?index=true. Accessed 16 May 2024.

Iovino, Serenella, and Serpil Oppermann, eds. 2014. *Material Ecocriticism*. Bloomington, Indiana University Press.

James, Erin. 2015. *The Storyworld Accord: Econarratology and Postcolonial Narratives*. Lincoln NE: University of Nebraska Press.

Johns-Putra, Adeline. 2019. Climate and History in the Anthropocene: Realist Narrative and the Framing of Time. In *Climate and Literature*, ed. Adeline Johns-Putra. Cambridge: Cambridge University Press.

Kerridge, Richard. 1998. Introduction. In *Writing the Environment: Ecocriticism and Literature*, ed. Richard Kerridge and Neil Sammells, 1–9. London: Zed Books.

Laverty, Megan. 2023. A Resolute Reading of *Metaphysics as a Guide to Morals*. *Iris Murdoch Society Blogpost*, February 9. https://irismurdochsociety.org.uk/2023/02/09/resolute-reading/. Accessed 9 May 2023.

Leeson, Miles. 2022. Is Murdoch a Philosophical Novelist? In *The Murdochian Mind*, eds. Silvia Caprioglio Panizza and Mark Hopwood, 363–375. London: Routledge.

Leeson, Miles, Megan Laverty, and Evgenia Mylonaki. 2023. Metaphysics as a Guide to Morals Podcast 2. *Iris Murdoch Podcast*, March.

Marland, Pippa. 2013. Ecocriticism. *Literature Compass* 10 (11): 846–868. https://www.researchgate.net/publication/263223901_Ecocriticism. Accessed 16 August 2021.

Morton, Timothy. 2007. *Ecology Without Nature: Rethinking Environmental Aesthetics*. Cambridge, MA: Harvard University Press.

———. 2018. *Being Ecological*. London: Pelican.

Nixon, Rob. 2011. *Slow Violence and the Environmentalism of the Poor*. London: Harvard University Press.

Nussbaum, Martha. 2015. *Political Emotions: Why Love Matters for Justice*. Cambridge MA: Harvard University Press.

O'Neil, Luke. 2018. Earthrise at 50: The Photo That Changed How We See Ourselves. *Guardian*, December 22. https://www.theguardian.com/science/2018/dec/21/earthrise-photo-at-50-apollo-8-mission-space-nasa. Accessed 18 July 2024.

Oak Taylor, Jesse. 2019. Atmosphere as Setting. In *Climate and Literature*, ed. Adeline Johns-Putra, 31–44. Cambridge: Cambridge University Press.

Oulton, Lucy. 2014. *'Being in the Presence of Beauty': Art, Affect and the Embodied Mind in Novels by E.M. Forster, Iris Murdoch and Zadie Smith*. Unpublished Masters Dissertation, Kingston University.

———. 2020. Loving by Instinct: Environmental Ethics in Iris Murdoch's *The Sovereignty of Good* and *Nuns and Soldiers*. Études britanniques contemporaines 59: *Iris Murdoch and the Ethical Imagination–Legacies and Innovations*. https://doi.org/10.4000/ebc.10237.

———. 2022. Nature and the Environment. In *The Murdochian Mind*, ed. Silvia Caprioglio Panizza and Mark Hopwood, 453–467. London: Routledge.

Popova, Maria. 2019. *Figuring*. Edinburgh: Canongate Books.

———. 2022. The Poetry of Science and Wonder as an Antidote to Self-Destruction: Rachel Carson's Magnificent 1952 National Book Award Acceptance Speech. *The Marginalian*, November 30. https://www.themarginalian.org/2022/11/30/rachel-carson-national-book-award-speech/. Accessed 14 October 2024.

Powers, Richard. 2018. *The Overstory*. London: William Heinemann.

Read, Daniel. 2019. *The Problem of Evil and the Fiction and Philosophy of Iris Murdoch*. Unpublished Thesis, Kingston University, June.

Rowe, Anne. 2002. *The Visual Arts and the Novels of Iris Murdoch*. Lampeter: Edwin Mellen Press.

———. 2019. *Iris Murdoch*, Writers and Their Work. Liverpool, Liverpool University Press.

Rueckert, W. 1996. Literature and Ecology: An Experiment in Ecocriticism. In *The Ecocriticism Reader: Landmarks in Literary Ecology*, ed. Cheryll Glotfelty and Harold Fromm, 105–123. Athens GA: University of Georgia Press.

Sagare, S. B. 2001. An Interview with Iris Murdoch. *Modern Fiction Studies* 47 (3): 696–714.

Soper, Kate. 1995. *What is Nature?: Culture, Politics and the Non-human*. Oxford: Wiley-Blackwell.

Stevenson, Angus, ed. 2010. *Oxford Dictionary of English*. 3rd edn (*OED*). Oxford: Oxford University Press.

Taylor, Craig. 2019. Fact and Value (*MGM* Chapter 2). In *Reading Iris Murdoch's Metaphysics as a Guide to Morals*, eds. Nora Hämäläinen and Gillian Dooley, 67–78. London: Palgrave Macmillan.

Uri, John. 2020. 90 Years of Our Changing Views of Earth. *NASA History*, December 21. https://www.nasa.gov/feature/90-years-of-our-changing-views-of-earth. Accessed 3 November 2021.

Vidal, John. 2012. Sellafield: "It Was All Contaminated: Milk, Chickens, the Golf Course". *Guardian*, March 11. https://www.theguardian.com/environment/2012/mar/11/sellafield-stories-book-nuclear-accident?intcmp=122. Accessed 25 April 2023.

Waugh, Patricia. 2012. Iris Murdoch and the Two Cultures. In *Iris Murdoch: Texts and Contexts*, ed. Anne Rowe and Avril Horner, 33–58. Basingstoke: Palgrave Macmillan.

Weik von Mossner, Alexa. 2018. From Nostalgic Longing to Solastalgic Distress: A Cognitive Approach to Love in the Anthropocene. In *Affective Ecocriticism: Emotion, Embodiment, Environment*, ed. Kyle Bladow and Jennifer Ladino. Lincoln NE: University of Nebraska Press.

Weil, Simone. 1952. *Gravity and Grace*. Translated by Emma Craufurd. London: Routledge.

Westling, Louise. 2011. Merleau-Ponty's Ecophenomenology. In *Ecocritical Theory: New European Approaches*, ed. Axel Goodbody and Kate Rigby, 126–138. Charlottesville, VA: University of Virginia Press.

White, Frances. 2019. The Gifford-Driven Genesis and Subliminal Stylistic Construction of Metaphysics as a Guide to Morals. In *Reading Iris Murdoch's Metaphysics as a Guide to Morals*, ed. Nora Hämäläinen and Gillian Dooley, 17–31. London: Palgrave Macmillan.

———. 2020. Anti-Nausea: Iris Murdoch and the Natural Goodness of the Natural World. *Études britanniques contemporaines* 59: *Iris Murdoch and the Ethical Imagination–Legacies and Innovations*. https://doi.org/10.4000/ebc.10212.

Williams, Raymond. 2019. *The Country and the City* (1973). London: Vintage.

Woods, Derek. 2014. Scale Critique for the Anthropocene. *The Minnesota Review* 83.

Woolf, Virginia. 1954. *A Writer's Diary*. London: Hogarth Press.

Zalasiewicz, J., et al. 2015. When Did the Anthropocene Begin? *Quaternary International* 383: 196–203. https://doi.org/10.1016/j.quaint.2014.11.045.

2

Iris Murdoch, Poet: Ecological Themes and Lyrical Influences

'THE PATIENT ARTIST STRIVING TO CONFINE
ALL NATURE'S LOVELINESS WITHIN A LINE'

At one time Iris Murdoch was intent on becoming a poet. She was already a keen versifier before the 'madnesses of Europe hurt [her] into moral philosophy' (*EM* xix).[1] On 18 December 1948, she wrote wistfully to French writer and critic Raymond Queneau, 'Imagining my life will begin when I chuck philosophy and just write novels and poetry' (*LoP* 116). However, when asked about poetry in an interview three decades later, her response was remarkably modest (as it turns out): 'Now I am a poet manqué. I have been writing verse all my life and I think I have probably written about eight poems. I mean I constantly beaver away writing poetry and trying to get it right', she tells Christopher Bigsby (*TCHF* 105). We have only recently come fully to appreciate what a truly

---

[1] In his preface to Murdoch's *Existentialists and Mystics*, Conradi adapts a line from Auden, 'mad Ireland hurt you into poetry' (*EM*, xix), to convey Murdoch's situation; her plans were upended after witnessing the devastating effects of war on ordinary people while working for the United Nations Relief and Rehabilitation Association (UNRRA).

prolific writer of poems Murdoch was following the discovery of her poetry notebooks in the attic at Charlbury Road, Oxford, and their subsequent arrival at the Iris Murdoch Collections at Kingston University Archives and Special Collections in September 2016. This important acquisition means that researchers are able, at last, to scrutinize how Murdoch honed her poetic art from a young age and her writing craft more generally. In these holograph drafts, she reveals the inspiration derived from an extensive knowledge of many great poets and an emerging ambition to develop her own poetic voice.

This extensive, first-hand, receptive and productive experience informs her view when Murdoch declares, in an interview with Bryan Magee in 1977, that poetry is the 'hardest kind of literature' (LP 5). In one important sense, Murdoch views poetry as akin to philosophy. Poetry like philosophy, she says, involves 'a special and difficult purification of one's statements, of thought emerging in language' (LP 5). However, in the ensuing discussion she describes to Magee how literature as a whole diverges from philosophy, alighting upon the 'sensuous nature of art [...] concerned with visual and auditory sensations and bodily sensations' as belonging to literature's distinctive characteristics (LP 10). These sensuous qualities are evident in the early poems as Murdoch begins to hone a sensibility that becomes more apparent later in extraordinarily lyrical passages in the novels. This chapter selects just a few of Murdoch's many poems and discusses them in the context of some of the poets from whom she drew inspiration, and with whom the selected poems thematically align. Ecological themes, often inconspicuous within the complex narratives of her widely acclaimed fiction, emerge early on in her literary career through this more personal and purified practice of versification.

Until the poetry notebooks arrived at the Iris Murdoch Collections, the only available books of Murdoch's poems were a small calendrical selection of short poems, *A Year of Birds*, and a short-run volume, *Poems by Iris Murdoch*, which gathers together the poems she published in journals and magazines during her lifetime. *Poems by Iris Murdoch*, compiled and introduced by Yozo Muroya and Paul Hullah, was published in Japan in a limited print-run of 500 copies in 1997. The fully authorized volume was nonetheless produced under certain strict conditions: Murdoch 'had absolute control and the final say', Hullah recalls (2020).[2] In a contemporaneous review of the collection, Geoffrey Heptonstall suggests that Murdoch 'came late to poetry', appearing to overlook the fact that all of the poems in *Poems by Iris Murdoch* had been previously published—almost half of them before she had turned twenty (1999, 84).[3] Importantly, however, Heptonstall also tells us that while he sees Murdoch's poems as situated at the margin of her oeuvre, they nonetheless 'may lead us to the centre' (1999, 89), thus acknowledging the poetry's potential to offer certain insight into the novels. Murdoch's early poems prioritize perception and emotion and, as Hullah observes, establish for her reader her lifelong 'fascination with the mysterious power of nature' (*PIM* 42). Themes emerge in Murdoch's early poetry that help draw attention to the ecological consciousness that eventually re-emerges in her novel-writing.

'The Spontaneous Overflow of Powerful Feelings'

Throughout much of Murdoch's writing career, she recorded poem after poem in her poetry notebooks but only a few of them were ever published. Mostly, her poetry-writing was a private endeavour. The influence of the Romantic poets is likely to have been born of her intimate knowledge of their works. A letter, sent to Oxford friend Hal Lidderdale by Murdoch on 6 November 1945 while she was working for the United Nations Relief and Rehabilitation Administration's (UNRRA) British Mission in Belgium, reveals a little of this knowledge. Evoking Blake's 'The Tyger', she asks Lidderdale, 'Where are you now, in what distant deeps or skies?'; in the same letter, she compares her enormous excitement at meeting Sartre—'*I have met Jean-Paul Sartre!* […] He is small, simple in manner, squints alarmingly and talks exquisitely'—to her

---

[2] This and further references to email correspondence with Paul Hullah on 28 January 2020 are reproduced by kind permission of the author.
[3] By 1939, Murdoch's poetry had already been published in *Badminton School* magazines 61, 69, 72 and 75; Bristol School Boys and Girls, *Poet Venturers: A Collection of Poems*, KUAS79; *The Cherwell* 56.3 (13 May 1939) and 57.4 (11 November 1939); and *Oxford Forward* 12 (19 May 1939); see, Appendix 4 (*PIM* 108–9).

childhood discovery of the great poets: 'I remember nothing like it since the days of discovering Keats and Shelley and Coleridge when I was very young!' (*LoP* 55). The Romantic influence—William Blake, William Wordsworth, Samuel Taylor Coleridge—asserts itself in Murdoch's apprehension of a stripped back elemental, even subliminal essence in her work, particularly in some of her early poems that relate to the natural world. In 'The Phoenix Hearted', published in *Badminton School* magazine shortly before she leaves the school in 1938, she articulates her yearning mission: 'The patient artist striving to confine / All nature's loveliness within a line' (*PIM* 54). Her work is starting to display the kind of enactive understanding that is often attributed to Romanticism:[4] the movement credited with discovering poetry's propensity for revealing and addressing 'fundamental truths of existence' and 'for capturing prelinguistic experiences and sensations' (Bradford 2010, 90). In the preface to his *Lyrical Ballads*, Wordsworth famously declares, 'All good poetry is the spontaneous overflow of powerful feelings: it takes its origin from emotion recollected in tranquility' (2007b, xxxiii). Experimenting with these kinds of powerful feelings became a Murdochian preoccupation and was destined, later on, to become a significant hallmark of her novels.

In *A Word Child*, two characters compare poetry to the novel in a way that reveals Murdoch's deeply held respect for the art of poetry. In response to Hilary Burde's claim that novels have the capacity to clarify life, Arthur Fisch attests to poetry's ultimate power: 'Poetry', he declares, 'is where words end' (*WC* 88). Hullah goes as far as to suggest that were one to consider this maxim an authorial position, then Arthur offers real insight into Murdoch's thinking on the subject. Hullah writes:

> It is a perspective which sees poetry as allowing a direct and spontaneous response to the natural world […]. It is a conviction that aligns its originator to the earliest notions of Romanticism and that movement's claims to have fused being and landscape in art as a way of comprehending existence. (*PIM* 30)

---

[4] Enactivism comes from cognitive science, whereby cognition arises from an 'actor' (an acting organism) interacting with their environment; see Catherine Read and Agnes Szokolszky (2020) 'Ecological Psychology and Enactivism: Perceptually-Guided Action vs. Sensation-Based Enaction', *Frontiers in Psychology* 11 https://doi.org/10.3389/fpsyg.2020.01270

However, Murdoch also saw great danger in a movement that was seen as promoting the poet of privileged means 'purveying a myth of individual inspiration' (Bate 1991, 45), and living a life removed from the materiality and hard labour of everyday rural life. In 'Against Dryness', Murdoch describes just such a 'lonely self-contained individual' as 'what is left of the other-worldliness of Romanticism when the "messy" humanitarian and revolutionary elements have spent their force' (AD 292). Murdoch is referring here, of course, to the political and economic fall-out of the French Revolution, and the social turbulence and radical politics of the Romantic period that followed. Some of Murdoch's British literary forebears (notably John Clare) will have witnessed rural dispossession, widespread industrialization and, with it, increasing urban migration. London, in particular, experienced a swelling population, and the associated environmental impact of all these developments would manifest soon enough (Hutchings 2007, 175). But writers of the Romantic period generally had the means and the education to be left fairly unaware of the difficult economic experiences of ordinary people, particularly the harsh living of increasingly depopulated rural communities. Thus, for Murdoch, Romanticism had its significant limitations and she remained circumspect about the movement's legacy. In her essay 'The Sovereignty of Good over other Concepts', she warns against the temptation of 'exalted self-feeling' (SGC 369), using attention to the natural world as her exemplar:

> It may seem odd to start the argument against what I have roughly labelled as 'Romanticism' by using the case of attention to nature. In fact I do not think that any of the great Romantics really believed that we receive but what we give and in our life alone does nature live, although the lesser ones tended to follow Kant's lead and use nature as an occasion for exalted self-feeling. (SGC 369)

Here, Murdoch has embedded a couplet from Coleridge's 1802 autobiographical poem 'Dejection: An Ode' (1997, 307–11, IV, l. 47–48), to then roundly reject the notion that either he or the other 'great Romantics' among his contemporaries could really have believed that the natural world has no life of its own (no animism) but can only be set alive with meaning by humankind. 'Dejection' contradicts the pantheistic position that Coleridge had already established in 'The Eolian Harp' (1795), for instance:

> O! the one Life within us and abroad,
> Which meets all motion and becomes its soul,
> A light in sound, a sound-like power in light,
> Rhythm in all thought, and joyance everywhere—(1997, 87–88, l. 26–29)

Here, the 'one Life' represents his conception of a deep connection between humanity and the natural world, 'Which meets all motion and becomes its soul'. In another of his poems, 'Frost at Midnight' (1798), Coleridge avers that the divinity of nature teaches spiritual perfection, thereby connecting the natural world to the Almighty; here he claims that nature even speaks the language of God:

> But thou, my babe! shalt wander like a breeze
> By lakes and sandy shores, beneath the crags
> Of ancient mountain, and beneath the clouds,
> Which image in their bulk both lakes and shores
> And mountain crags: so shalt thou see and hear
> The lovely shapes and sounds intelligible
> Of that eternal language, which thy God
> Utters, who from eternity doth teach
> Himself in all, and all things in himself. (1997, 231–33, l. 58–61)

By embedding Coleridge's line into her argument in this way, Murdoch clearly signals that she does not accept the sentiment expressed in 'Dejection' as suitably indicative of Coleridge's true position on the natural world's animism (as we might call it today); he is not one of those accused by Murdoch of following Kant in this regard. Instead, she numbers him among the greats who attended to the world, the natural world in all its particularity, in a manner that rises above the sort of Romanticism she decries: the sort of Romanticism that is guilty of fomenting dangerous self-obsession.

During the last thirty years or so, there has been a drive to reassess the Romantic movement in light of growing environmental concern. The particular interest lies in how it can be argued that the Romantic poets teach us to look at the natural world. Bate is one critic who has encouraged his readers to re-examine Wordsworth's writings for their significant ecological interest. He suggests that Wordsworth, for example, is struck by the sublime qualities of the natural world and affords it moral consideration. Wordsworth's philosophical thought is conveyed succinctly through a short excerpt from his lengthy poem fragment 'The Wanderer' in *The Excursion* Book 1, when

> An *active* principle:—howe'er removed
> From sense and observation, it subsists
> In all things, in all natures; in the stars
> Of azure heaven, the unenduring clouds,
> In flower and tree, in every pebbly stone
> That paves the brooks, the stationary rocks,
> The moving waters, and the invisible air. (2007a, 276, l. 3–9)

Wordsworth's speaker, the inveterate Wanderer living off the land, sees animism (the '*active* principle') in everything around him. His bookish creator may be enthralled by the natural world but is a step or two removed from participation in, or a life so directly influenced by, the natural world's vagaries. For Bate, the innate relationship between man and natural environment that Wordsworth is able to conjure in his Wanderer is very significant. He explains, 'the passion and the wandering are Wordsworth's, but the Wanderer is a Wordsworth without the influence of books and a Cambridge education' (Bate 1991, 65). Of course, reverting to an uneducated, and to a sense of an unmediated, awareness of nature presents a real challenge for the poet. Wordsworth has also been criticized (in a more hierarchical age than today) for expecting his audience to attend to the musings of a naïve wanderer who is also peddling wares (Bate 1991, 65, 119n). John Ruskin, on the other hand, saw Wordsworth's Wanderer as the purveyor of an 'ideal of vision': providing an opportunity to view the natural world through new eyes (Bate 1991,

65). Whatever the potential for derogation or otherwise, Wordsworth's depiction of the Wanderer's way of life proved sufficiently unorthodox to Wordsworth's readership (likely both monied and educated 'given [*The Excursion*'s] length, its price, and its abstract diction') to have been made all the more compelling at the time (Bate 1991, 65).

Bate offers an account of the Wordsworthian philosophy that underscores his poetry: the 'animation in and unity between all things' (Bate 1991, 65). A philosophy that is naturally self-evident to someone like the Wanderer should unquestionably mean that 'nature is accordingly entitled to moral consideration' by Wordsworth himself (Bate 1991, 65). In 1909, critic Leslie Stephen suggested that '[n]owhere is it easier to observe the mode in which poetry and philosophy spring from the same root and owe their excellence to the same intellectual powers' (Bate 1991, 64, cites Stephen 1909, 255). Stephen's comments are of course directed at Wordsworth, but there is an undeniably familiar ring to these words for scholars of Murdoch.

Murdoch's technique in her own poems, many written some years before she became known as a novelist, is to register the natural world affectively through the human body she sees as being a part of nature. She recognizes the separate existence of the world from self, but she knows that that same self tends to intervene to obstruct true vision. At the same time, some of these poems acknowledge planetary indifference to human affairs. Murdoch's appreciation of the 'independent existence' of the natural world, of 'animals, birds, stones and trees' (SGC 369–70), grounds the poems featured in this book in a broader ethics of place. The major environmental themes at play in her poetry include expressions of shifting temporalities, the integral and co-reliant involvement of human and environment, environmental desecration by human hand that co-exists with some sort of faith in the regenerative power of the Earth, and an ethics of care for all life including the smallest of creatures. Later, Murdoch will carry all of these elements forward into her novel writing.

'I Wish I Could Write Poetry'

Murdoch's early poems affirm a brilliant mind with wide-ranging interests. They reveal her powers of observation: the world's complexities and its intricate details. She had published eleven poems by the age of nineteen, some in the *Badminton School* magazine, and others in a fundraising booklet. The soft-bound anthology *Poet Venturers*, conceived by Murdoch and produced with the help of publisher (and parent at the school) Victor Gollancz in 1938, raised money for the Fund for Chinese Medical Aid.[5] Murdoch persuaded W.H. Auden, visiting Badminton School at the time of the anthology's compilation, to write the foreword (*IMAL* 78). She found so much to admire about Auden—'a truly *great* poet' (*LoP* 488)— and he was set to become a life-long source of inspiration to her. On the endpaper of her copy of *The Enchafèd Flood* she has inscribed the words, 'revisiting the source of one's inspiration. As you are one, I am revisiting you' (IML 81). The four poems by Murdoch in *Poet Venturers*, 'The Phoenix-Hearted', 'Star-Fisher', 'The Coming of April' and 'Lower than the Angels', were subsequently republished alongside the work of Samuel Pepys, Alexander Pope, Coleridge, Wordsworth, Auden and John Betjeman in a volume titled *600 Years of Bristol Poetry* in 1973.[6]

Of 'The Phoenix Hearted', Priscilla Martin and Anne Rowe have noted an 'obvious' Yeatsian influence (2010, 3). The poem is certainly an early expression of the political Murdoch and, given her likely source of inspiration, arguably also acknowledges her Irish heritage. The Second Sino-Japanese War (1937–1945) had been raging for a year when Murdoch wrote 'The Phoenix Hearted' and, like Yeats's 'Lapis Lazuli', her poem speaks to the fundamental human need for the arts: their power to inspire, motivate and harness regeneration and renewal in the aftermath of war and destruction. Echoing Yeats's 'All things fall and are built again / And those that build them again are gay (2001, 152, l. 35–36), Murdoch's poem ends with its own glimmer of optimism:

---

[5] Bristol School Boys and Girls, *Poet Venturers: A Collection of Poems* (1938), KUAS79, in the Iris Murdoch Collections.
[6] See Edward Martin and Bill Pickard (eds.) (1973, 49–51). Photocopies of these pages are located under KUAS6/10/2/7 in the Peter J. Conradi Archive at Kingston University Archives and Special Collections.

> But strength in stillness lies. She is not dead.
> Kuan Yin again shall raise her gracious head,
> And though it fall in flames of bloody red
> The phoenix shall arise with plumage proud.⁷

Poet Venturers marked the triumphant end of Murdoch's school career. By the time she left Badminton School, she bore little resemblance to the sad, shy child who had arrived in 1932 and had to be encouraged to help in the school greenhouse to distract her from 'terrible homesickness' (*MA* 15).

The poems Murdoch succeeded in publishing from 1939 onwards, together with the many others she composed in the poetry notebooks she kept during her university years and beyond, establish a substantive and engaging prequel to the story of her subsequent fame as a respected philosopher, public intellectual and acclaimed novelist. In 1939 with Murdoch now at Oxford University, the *Cherwell* published 'Poem' and 'Untitled II' and the *Oxford Forward* published 'Oxford Lament'.⁸ At the time, strategically I like to imagine, Murdoch found herself on the staff of both papers (*MA* 22). The *Cherwell* ('Oxford's oldest student newspaper') had already championed the work of a string of undergraduate-and-destined-to-be-illustrious figures: Evelyn Waugh, Graham Greene, John Betjeman, L.P. Hartley, C. Day-Lewis and Auden (Walmsley 2016). Murdoch's poetry notebooks reveal that some of her most productive years were during this period at Oxford. It would seem quite wrong to disregard this early work as immature experimentation, as juvenilia, when this emerging creative talent was set for recognition as one of the most influential British novelists of the twentieth century. On the contrary, this work demands our attention.

Hullah has said of Murdoch's poetry that 'she starts off clearly inhabit[ing] the style of other poets, aping set forms with great aplomb', but 'by the time we reach the later poems she is finding her very own

---

⁷ Iris Murdoch, 'The Phoenix Hearted', first published in *Badminton School* magazine no. 77, Summer Term (1938, 18, l. 27–30), is reproduced in the two anthologies *Poet Venturers* (KUAS79, 5) and *Poems by Iris Murdoch (PIM* 54), and in Martin and Rowe (2010, 1).
⁸ Iris Murdoch, 'Poem', *Cherwell* 56.3, 13 May 1939 'Untitled II', *Cherwell* 57.4, 11 November 1939; 'Oxford Lament', *Oxford Forward* 12, 19 May 1939.

style' (2020). He describes how, once all of Murdoch's poetry had joined the Iris Murdoch Collections, he was 'so uplifted finally to read the unpublished notebooks' because, in his view, there is 'conclusive evidence there that she is a notable, original poet':

> The meagre selection in [*Poems by Iris Murdoch*] was [...] inconclusive evidence of that, I think. A leap of faith was required at that juncture that, once one sees all the poems we now have that we didn't back then, is no longer necessary. (Hullah 2020)

Evidence gathered from the journals, the poetry notebooks and *Poems by Iris Murdoch* suggests that Murdoch was unable to lend so much time to her poetry after leaving university. Indeed, thirty-six years elapse before she publishes further poems, entering a particularly productive phase in the 1970s when, once more, she appears able or inclined to attend to this aspect of her work. From 1975 and over the ensuing fifteen-year period, seventeen poems appeared in publications as diverse as the *Boston University Journal*, *The Listener*, *Harpers & Queen*, and Susan Hill's *People: An Anthology* (*PIM* 108–9).

The ten poetry notebooks, together amounting to a vast number of poems, suggest a certain ambition, an ambition, however, that is tempered with a deep-seated insecurity.[9] On 27 December 1958, Murdoch pens a note to her editor, Norah Smallwood, at Chatto & Windus:

> I write in a mad moment to ask the following. I have by me a small number of poems, written at various times. They are obviously not very good. However, they exist and occasionally occupy my thought [*sic*]. John [Bayley] likes a few of them. It sometimes occurs to me that I would like to shew them to someone else, not necessarily with any view to publication. (*LoP* 192)

In 1959, on abandoning 'Jerusalem' (a novel, never published and whereabouts unknown), Murdoch notes in her journal that she finds her solace in poetry-writing is laced with a certain frustration:

---

[9] The ten poetry notebooks (known as Poetry Notebooks 4 to 13) are located at KUAS202/3/4–13 in the Iris Murdoch Collections.

> I have written a lot of poetry lately & really don't want to do anything but that. It is not totally bad poetry but mediocre. I wish I could write poetry. Extraordinary inward agitation with no apparent cause. At a time like this I <u>see</u> & <u>know</u> that all art is ultimately love. But this knowledge lies dark in me & I have not the <u>courage</u> & the <u>goodness</u> required to do anything with it. (J9 35)

To compound the intrigue, later that same day she adds with no further qualification: 'In the poetry I am trying to change my old wicked god into a good god. But to change our gods we have to change ourselves' (J9 35). Her soul-searching insecurities and self-reflexive remarks were not, however, limited to poetry. On 10 January 1971, she records in her journal the '[f]irst aconite', before noting down questions that are clearly troubling her, together with her own responses:

> Why did I write no verse this autumn? The novel [*The Black Prince*] made it impossible, I could not give anything elsewhere. The novel is too intellectual, it is all getting too intellectual. Break it up. But how? TSTS [her play, The Servants and The Snow] was too intellectual. (J11 47)

'The Servants and the Snow' marked a shift from adapting novels for theatre to 'writing directly for the stage where political themes could be explored freely' and, moreover, represented 'an attempt to reach a wider audience than that which she envisaged habitually read her novels', explains Rowe (*IMR*12 40). The play was produced alongside an explanatory essay written by Murdoch for Greenwich Theatre's magazine *Cue*, the very existence of which perhaps suggests a dramatic complexity that some audience members may have been unprepared for. Drama, she declared, is 'a public form of poetry' (*Cue* 13–14, cited in *LoP* 356). The play closed after four weeks and Murdoch was very distressed by its failure (*IMAL* 531). *The Black Prince*, the novel that Murdoch worried was also 'too intellectual' (J11 47), was written during an intense four-and-a-half month period the following year. In fact, it was not only well received on publication in 1973, but went on to win the James Tait Black Memorial prize and to reach the short-list for the Booker prize later the same year.

Five years later, at the height of her powers as a public intellectual, Murdoch senses 'more confidence in my ability to write plays, poetry'

(J12 62). However, her insecurity lingers in relation to a particular selection of poems she is preparing for publication. She appears to have spent a prodigious amount of time revising a collection of poems that in the event was never published, titled 'Conversations with a Prince'.[10] Such conscientious revisioning, when poems are worked, reworked and worked again long after the inceptive moment of inspiration, risks conflicting with 'our desire, our elemental need to understand the poem is instinctive', a quality, according to Richard Bradford, that is as fundamental to a poem as its meaning (Bradford 2010, 258). An element of spontaneity, itself a product of the initial inspiration, must be at risk of getting lost in such a thorough revisioning process. In 'Thinking and Language', Murdoch ascribes the specific task to poetry of representing the 'give and take' of language: 'words may determine a sense, or a fresh experience may renew words' (TL 36), which lends some justification to her process. Nevertheless, some of the less worked-up and less self-conscious poems (both published and unpublished) of the younger Murdoch are those that seem more readily to reveal her ecological vision, through their organic depictions. They represent the sort of vision that, carried forward, re-emerges in the novels but finds itself competing thematically with a great many others.

It is important to emphasise that my purpose here is not to evaluate Murdoch's poetic output and interrogate her status as a poet among poets even though Hullah, a poet himself, describes her poetry as 'really rather good: suggestive and surprising, rhetorically rich, and technically adept' (2023, 67). Hullah thinks her poetry should be viewed as 'an illuminating companion text to all her prose' (2023, 67). Indeed, Murdoch chose to leave most of her poems packed away undisturbed during her lifetime, their last resting place a box in the attic at her home in Charlbury Road (Rowe 2019, 94). Instead, with the natural world as the focus of this book, my purpose is to evaluate the extent to which some of Murdoch's poems reveal an early affinity, a sense of moral obligation, and, more generally, deliver expressions of her environmental imagination to her reader.

---

[10] See Murdoch, 'Conversations with a Prince', KUAS202/3/1–3 in the Iris Murdoch Collections.

The Nature of Poetry

It is useful, I think, to unpick some of the terminology relating to the sort of poetry that concerns itself with the natural world. Introducing the second edition of his book *Green Voices*, Terry Gifford explains that one of the greatest challenges that confronts contemporary poets, when choosing the natural world as their subject, is the issue of whether one can call such poetry 'nature poetry' (2011, 7–24). Publishing his first edition, Gifford predicted that 'green poetry' would become the more widely used term for this sort of work on the basis of all-encompassing inclusivity—hence his choice of title for the volume. Instead, what used to be known as 'nature poetry' now carries a rather more homiletic term: 'ecopoetry' (2011, 8). 'Green poetry' has become the term for the narrow art of 'propagandist environmental poetry', Gifford asserts, 'nature poetry' having already become a 'pejorative designation' (2011, 7). He flatly rejects Bate's claim that a poem fails to be ecological when the 'language itself is not being asked to do ecological work' (Gifford 2011, 8). While he can accept the immortal 'poetry makes nothing happen'—invoking Auden's tribute poem 'In Memory of W.B. Yeats' (1979, 81, II l. 5)—what Gifford says he really demands of poetry, with the natural world as its focus, is the potential for a 'shift in sensibility' (2011, 8). This shift may, of course, be what Auden intends with 'it survives / A way of happening, a mouth' (1979, 81, II l. 9–10). Susanna Lidström and Greg Garrard draw a helpful distinction between 'ecophenomenological poetry', which they define as work that 'focus[es] on descriptions and appreciation of non-human nature with roots in Romantic and deep ecology traditions, aiming to heighten individual readers' awareness of their natural surroundings', and 'environmental poetry', which they see as attempting to 'grapple with the changing relationship between human societies and natural environments' (2014, 37). Murdoch's work most obviously belongs to the first of these categories. Evelyn Reilly asserts in broader-brush fashion that 'ecopoetics' is not a poetic style or movement at all, but simply the 'fact of writing in the world of accelerated environmental change' (2013). Reilly's perspective on poetry, ecology and climate neatly aligns with Bavidge's thoughts on the novel discussed at the start of this book. It seems, regardless of genre, environmental concerns frequently lurk in the minds of readers and cannot fail to surface when reading landscape, habitat or environment. According to Scott Bryson, ecopoetry is 'heavily dependent upon ideas of ecology and viewing things in concert [and] not in isolation' and comprises three elements: a recognition of the 'interdependent nature of the world' (ecocentrism), 'an imperative toward humility in relationships with both humans and non-humans' and an 'intense skepticism concerning hyperrationality' (2002, 7). Modernity's overreliance on technology, he believes, builds a tendency to this excessive rationality and, without a good level of scepticism, our 'overtechnologized modern world' is in danger of inducing ecological disaster (Bryson 2002, 6). His position clearly resonates with Murdoch's concerns, expressed decades earlier, about the creep of scientism.

John Felstiner argues, in his bleakly titled *Can Poetry Save the Earth?*, for the profound value of ecopoetry. He asserts that we 'sense but can't really grasp stone or tree, let alone stream or bird. Still, at times, the saving grace of attentiveness, and the way poems hold things still for a moment, make us mindful of fragile resilient life' (Felstiner 2009, 357). The Murdoch poems included in this chapter uphold this 'saving grace of attentiveness', an attentiveness (or attention) that represents the centrepiece of her moral philosophy. How hard it is to adhere to a high level of attentiveness to the other is made manifest, Murdoch argues, in the general tendencies of human behaviour. She says, 'so much of human conduct is moved by mechanical energy of an egocentric kind. In the moral life the enemy is the fat relentless ego' (OGG 342). For Murdoch, the suppression of ego (unselfing) and achieving this high level of selfless attention represents something to aspire to, to work constantly towards. Murdoch's poems apprehend the rights and values that are integral to ecological processes and vital for non-human creatures, and their meaning is enriched in interpretation through a contemporary ecocritical lens. The key defining element of all the Murdoch poems discussed in *Iris Murdoch's Wild Imagination* is the conception of the living world's interconnections or connectedness.

The status and classification of contemporary 'nature poetry' ought perhaps not to engender too lengthy a discussion when Murdoch's poems, despite only having come to light in recent years, predate much of this thinking. What is important is the sense one gains of her holistic view of natural Earth's systems and how her ideas anticipate contemporary poems on the natural world. It is in the context of the connection subsisting in the Wordsworthian, 'In all things, in all natures' (2007a, 276, l. 5), or in the words of Murdoch's contemporary, American cellular biologist and once a leading ecologist Barry Commoner, in 'everything [being] connected to everything else' (1971, 33), that a close reading of her poetry becomes fruitful.[11] Kevin Hutchings explains the bald implications of any systemic breakdown of the planet: 'According to this holistic model of nature, if we interfere with or change any single part of a given habitat, we will introduce a ripple effect that inadvertently transforms the whole habitat' (2007, 176). Murdoch appears to share this conception of Earth's holistic processes, the co-reliance of all life, and an apprehension of the impending threat of environmental breakdown.

Thematically, Murdoch's poems featured in this book become more relevant as we address the urgent issues around the current climate crisis. While poetic practice in general tends not to address momentous events specifically, poetry nonetheless carries with it both historical and political significance according to Paul O'Flinn, by being envisioned against a 'cluster of tensions, hopes and fears' of the contemporary moment (2001, 9). O'Flinn's assertion is problematized when much of Murdoch's poetry only now comes under close critical scrutiny, three quarters of a century after it was first composed. The earliest poem selected for discussion in this study was written in 1938, in the tense and unsettling period just before the outbreak of the Second World War. Murdoch's nature poetry, specifically, is considered for the first time, inevitably scrutinized through a lens tainted by one's own 'cluster of tensions, hopes and fears' in the twenty-first century (O'Flinn 2001, 9) regarding ecological Earth, and at a time when climate breakdown is (or should be) firmly at the top of the international political agenda. In our environmentally stricken age, the prescient nature of certain of Murdoch's poems garners attention.

---

[11] An American cellular biologist and a leading ecologist, Commoner met Murdoch as a fellow member of the 'Study Group on Foundations of Cultural Unity', the interdisciplinary panel organized by Marjorie Grene, Michael Polanyi and Edward Pols, at Bowdoin College, Brunswick, Maine, 21–27 August 1966 (Grene 1969, ix–x).

Some Poems by Iris Murdoch

Murdoch's untitled poem recorded in her poetry notebook on 23 August 1943, beginning 'Gently I have touched the thin lids of your eyes', evokes erotic intimacy in the haptic memory of the body. In foregrounding bodily sensation, the tone is set for an intimate piece in which her speaker intuits both body and world. 'Gently I have touched' speaks of an obsessive love felt through the body, establishing a major affective theme in Murdoch's nature poems that re-emerges in the later novels.[12] One discerns the narrow shafts of light on a sun-drenched, tender and visceral scene. Light has a vivifying effect on both body and kingdom, and fingers and eyes remind us that the exploration of landscape is only possible through such powers of perception.

[Untitled]
Gently I have touched the thin lids of your eyes   1
In fear with golden fingers—and each limb
Has loved the ambushed sunlight's swift surprise,
Darting from warm wings of the seraphim.
Our bodies' kingdom had its intricate   5
And ancient laws—& the shafted light
Was over all, the cloud, the delicate
Thin air, thought's atmosphere, transparently bright.
I have another earth now & a new sky,
Driven by the angel out from the gate.   10
Infinite & cool with dew there lie
The world's lands for me. But no rose
Of all their roses can make me forget,
Forget how green, how green, that garden was.   14
23/8/43
(PN5 226)[13]

---

[12] All the poems by Murdoch in this chapter are reproduced by kind permission of Kingston University Archives and Special Collections; where a poem is untitled, it is referred to in the text by its first line.

[13] This is the final of four versions of the poem recorded in the notebook. Each version of the poem opens differently: 'Gently I have touched the splendour of your eyes' (PN5 141), 'Gently I have touched the sensitive lids of your eyes' (PN5 146), and 'The forest I remember & that sky' (PN5 149).

Murdoch engages her perceptive powers to deliver the first two quatrains of an imperfect sonnet; the turn comes early and there is no concluding couplet. She does not allow rhyme to interfere with syntax. Instead, she capitalizes on a change of tone, leaving the final lines unrhymed. The conjunctive turn in her sonnet, heralded by the caesura in the antepenultimate line, presents a change in modulation from passionate obsession to calm reflection. The composite form of this poem conveys meaning, then, when it frees itself from the constraints of the sonnet as the speaker has freed and sobered herself from love's obsession in order to reflect on its experience. The tone transitions from guileless love and sensory obsession to reflection on things past and future, the rich memory of her love. Murdoch demonstrates a capacity to produce a powerfully erotic poem to express the fall from innocence to experience, recalling Romantic poet and printmaker William Blake. She aligns seduction with the celestial, 'Darting from warm wings of the seraphim' (l. 4), and her 'no Rose / Of all their roses' (l. 12–13) speaks to the intimate exchange that recalls Blake's 'crimson joy' from his poem 'The Sick Rose' (1971a, 159, l. 6). She connects touch's sensory potential to a concomitant intermingling of the rich imagery of kingdom, lands, new earth and new sky. In this way a glintingly warm and mesmeric poem functions to obscure or blur the distinction between body and world. The fragility of life and the transient nature of such intimacy are set against the primal metaphors of bodies and worlds. The final line's verdant imagery conveys a health-giving richness to the experience, achieved through reification of the unnamed, ungendered lover as garden, 'The world's lands' (l. 12) suggesting future loves. The poem, while evoking the traditional Romantic nature lyric, presents an early and powerful invocation of the sensory nature of embodied interaction with landscape that she will go on to portray in many of her novels, including key scenes in the garden of Trescombe House, Dorset, in *The Nice and the Good* (1968), another in the Provençal foothills of Les Alpilles in *Nuns and Soldiers* and in the woodland beyond Seegard in *The Good Apprentice*.

[Untitled]
The morning fills my eyes & my heart, 1
The brown earth rises up to my feet

> Caressingly. The white wind touches my breast
> Tenderly. Let me live now
> Only in nature's larger life—& be   5
> A curve of the hills, a breath of the wind,
> A wave of the sea.
> 4/2/38
> (PN4 12)

Murdoch adopts a similar technique in 'The morning fills my eyes & my heart', which is about connection—the human and the more-than-human seen as belonging within a single ecosphere: the living world's co-dependent, self-sustaining but potentially vulnerable system. For a brief moment humanity and the natural world are imagined as indistinct from each other. 'The morning fills' displays not only spatial but temporal awareness. The fragility and fleeting nature of human existence is set in the context of the 'larger life' (l. 5) of the elements, the sea and the land—an idea that even now seems little understood in this age of climate breakdown—and a theme that Murdoch will carry into the novels. It is not the existence of the planet itself that is under threat, but life upon it. 'The morning fills' is a hymn to precious Earth and the long-lived enduring presence of its materiality. And here the speaker is making a plea for temporal perspective, acknowledging 'nature's larger life' (l. 5) and the concomitant evanescence of human existence.

> **[Untitled]**
> What does it matter?   1
> Our microscopic melancholies,
> Our infinitesimal impatience?
> The universe is not interested. It
> Moves on.   5
> 8/2/38
> (PN4 14)

The admonitory tone of this abrupt short verse written just four days later, 'What does it matter?', presents the speaker's interpretation of a chilling indifference of the cosmos to the human condition. The insignificance of the many and intense petty worries of the self-obsessed individual—or arguably and equally the self-obsessed government of peoples—is expressed in the multisyllabic alliterative phrases of the second and third lines, 'microscopic melancholies' (l. 2) and 'infinitesimal impatience' (l. 3). The enjambement and abrupt final half-line serve to emphasize planetary disinterest, leaving the reader with a chilling picture of utter desolation. This simple five-line poem, remarkable in its concision and prescience, provides a bleak and early Murdochian assessment of human insignificance: our fleeting presence on a planet and in a universe quite indifferent to us. The poem bleakly suggests, regardless of our petty rivalries and the lack of human attention to its care, that Earth in some form will proceed without us.

Spatial and temporal perspectives provide the focus of this next poem, too. 'The trailing stars tell of dooms', published in the *Iris Murdoch Review* 5 in 2014, is one of the many love poems dedicated to Wallace Robson in 1952. Robson was an English Fellow at Lincoln College, Oxford, with whom Murdoch had 'a turbulent romance' in the 1950s, 'culminating in a broken engagement which caused much pain on both sides', according to Frances White (*IMR*5 8).

> **[Untitled] 'for WR 1952'**
> The trailing stars tell of dooms
> In a universe next door to ours.
> I have seen the fall of the world
> Poised at the intricate centre of flowers.
> Pretty one, pretty one, I say   5
> To the timid suspense of a cat –
> Profound in her enormous eye
> A powdery lamp is lit.
> Day comes like a settling bird
> That I coax to my window sill –   10
> Reality awaits the word
> That shall shatter it once & for all.
> What a tremulous structure it is,
> Focussed, suspended in place
> By the random congeries,   15
> The atomic form of the face.
> Let the personality list

A fraction out of its sense
And the shadows of particles
Will fall with a difference.   20
Will fall to create new things
And, the colour structure broken,
A new born planet sings
That the word has at last been spoken.
(PN6 12; *IMR*5 13)

The poem demonstrates what Robert Macfarlane calls a 'shift in time scheme', a quality of nature that 'flare[s] into futurity, as well as reverberating out of the past' (2007, 316). The speaker's gaze reaches into the profound depths of a cat's 'enormous eye', into the 'intricate centre of flowers' (l. 4), and out to the 'trailing stars' which 'tell of dooms / In a universe next door to ours' (l. 1–2). White points to an overworked or 'clichéd' use of imagery in the collection of poems that Murdoch dedicated to Robson (*IMR*5 8). She says that many of the verses 'are visceral, dashed rapidly down in raw emotional states; rough drafts, unpolished' (*IMR*5 8). However, 'The trailing stars' anticipates the attention Murdoch affords the kestrel in 'The Sovereignty of Good over other Concepts' (SGC 369) with the call to 'Let the personality list / A fraction out of its sense' (l. 17–18), so that 'the shadows of particles' fall to 'create new things' (l. 19, 21). An important factor in fully appreciating what 'The trailing stars' is able to do is to consider the ecocritical principle of looking past the imagery to the thing itself.

In Murdoch's novel *The Message to the Planet* (1989), Marcus Vallar, speaking to Alfred Ludens, asserts the importance of drawing on the truth of the thing in itself: 'One may say too that a description of the rose means nothing unless, as in poetry, it can *be* the rose' (*MP* 381). Marcus and Ludens are reflecting on Marcus's capacity to acquire a following. Alluding to the seventeenth-century poet, theologian and doctor Angelus Silesius's '*Die Rose ist ohne Warum./ Sie blühet, weil Sie blühet*' ['The rose blooms and knows not why'] (2011, 71 §289), Marcus expresses his belief that matters cannot always be explained or clarified.[14] The concen-

---

[14] My translation (literally: the rose is without reason, it blooms because it blooms).

tration camps of wartime Europe lie beyond reason for Marcus, but his overarching point is that, when something extends beyond explanation, imagery simply will not do. It is vital that what is seen is the truth of the thing in itself. For the speaker of Murdoch's 'The trailing stars', truth lies in the anguish of life's precarity; she reminds herself (and her reader) of the insignificance of our petty concerns and of human endeavour in the wider context of time and space: 'the shadows of particles / Will fall with a difference. Will fall to create new things' (l. 19–21). The poem suggests the sort of complex temporal layering that is to become a distinctive feature of Murdoch's later novels.

Murdoch's early poems express ethical attention to the smallest of Earth's creatures, a particular feature, too, of the later novels. The untitled poem that begins 'Brilliant oaktrees, gold mixed with their green' (l. 1) delivers its first celebratory lines in tribute to a bountiful harvest where 'earth rides & spills / In slopes behind the splendid suns' (l. 3–4).

> **[Untitled]**
> Brilliant oaktrees, gold mixed with their green
> Clutch at the heart of Sorrell-dusted hills.
> Heavy with harvest now, earth rides & spills
> In slopes behind the splendid suns. Unseen
> A million creatures tremble on a keen   5
> Edge of survival—In the earth and night
> Or saturated in the soaring light.
> Life's fierce volition wears a front serene.
> Here I with closed eyes lip a precipice.
> Life tinkles all its perishable glass.   10
> The warm stars rain upon me with a hiss,
> And golden suns to black eclipses pass.
> The hillside quivers at the wheels of Dis,
> And Proserpina trembles in the grass.
> 1/7/41
> (PN5 43)

The joy juxtaposes the conditions for insect life contributing to the health of the soil and Earth's bounty, and yet 'Unseen / A million creatures tremble on a keen / Edge of survival' (l. 4–5). A serenity conceals the fragility as if of 'perishable glass' (l. 10) of more-than-human life at the mercy of 'Life's fierce volition' (l. 8). Harvest is underway. The 'earth rides and spills' (l. 3) as crops are brought in with little thought for insects or other invertebrates, 'In the earth and night' (l. 6), involuntarily dug up out of the soil and 'saturated in the soaring light' (l. 7). Yet, by invoking the brutal rape of Proserpina at the hands of Roman god of the underworld Dis Pater, Murdoch's speaker expresses deep concern about the environmental consequences of not giving due consideration to the violated habitats of these small creatures.

Murdoch appears to derive her ideas for this poem from an amalgam of sources. Blake offers a conception of co-dependence—'Every-thing that lives, / Lives not alone, not for itself' (1971b, Plate 3 l. 69–70)—in his Book of Thel. Blake gives expression to it when 'The weak worm from its lowly bed' is summoned by the cloud and the 'clod of clay' (Plate 4, l. 81), on hearing the voice of this 'image of weakness'—so called by the timid girl, Thel—(l. 76), declares 'O beauty of the vales of Har! we live not for ourselves' (l. 84).[15] The work of the eighteenth-century parson-naturalist Gilbert White was also known to Murdoch. In *The Sea, The Sea*, Charles Arrowby declares that 'I would certainly like to write some sustained account of my surroundings, its flora and fauna. This could be of some interest, if I persevered, even though I am no White of Selborne' (TSTS 2). The 'chain of nature' is a conception that White was alert to; he wrote the following in his volume on the natural life around where he lived in the English county of Hampshire, *The Natural History of Selborne*:

> the most insignificant insects and reptiles are of much more consequence, and have much more influence in the œconomy of nature, than the incurious are aware of; and are mighty in their effect, from their minuteness, which renders them less an object of attention; and from their numbers and fecundity. Earth-worms, though in appearance a small and despicable link in the chain of nature, yet, if lost, would make a lamentable chasm. (White 1853, 151–52)

---

[15] For more on Murdoch's indebtedness to Blake, see Daniel Read, 'The Problem of Evil and the Fiction and Philosophy of Iris Murdoch' (unpublished thesis 2019, 234–300).

In Murdoch's 'Brilliant oaktrees', the poem's speaker warns of the weaknesses in this chain in the face of careless human activity and unfettered technological progress.

[Untitled]
Snowdrops smell
Like a Dior scent
Out of the mud
Miracles. 4
(PN8 13)

For many poets, the arrival of spring engenders optimism with the concomitant hope of renewal. In this next short and untitled poem of the early 1970s that begins 'Snowdrops smell', Murdoch expresses how remarkable it should be that such beautiful, delicately scented buds emerge 'Out of the mud'(l. 3). Her use of the word 'Miracles' (l. 4) speaks to an unceasing amazement at rebirth but also ironizes the inadequacy of the faux equivalence often made between pure, natural scents and factory-made perfumes. The miracle of regeneration recurs in 'Gunnera', a poem first published in 1979 in *Poetry London / Apple Magazine* and republished in the Muroya and Hullah anthology.

Gunnera
A bulky thing I buried in December
Like a huge fudge with tentacles upon
Or a dead puppy dog
On a dim yellow day as I remember
Digging the stiff clay like a funeral 5
Cringed in the freezing fog
Has lifted up fantastical in May
A vast rain forest dome
Of crinkled dew-bedizened leaves a grove
Out of that wet and admittedly weighty blob 10
Defeat of gravitation
By genial mix of seed and sod
And formal demonstration of

> The accuracy of God
> His green grotesque imagination.   15
> (PN8 148)

Murdoch has recorded a number of different versions of 'Gunnera' as she works to refine the poem.[16] The curiously strange metaphor of the 'huge fudge with tentacles upon' persists throughout the reworking. To the side of one version of the poem Murdoch notes as an alternative to the 'dead puppy dog' 'a plum pudding', but the 'puppy dog' prevails; on the same page she makes a note to herself to 'rhyme funeral' but does not appear to have attempted this seemingly impossible task (PN8 26).

For Hullah, 'Gunnera' represents a 'slightly sinister' poem that 'offers a semi-pantheistic appraisal of natural instances' that 'combine in nature to denote the unknowable power behind the "seed and sod" (l. 12) we sometimes decide to call God' (*PIM* 40). The poem concerns the December planting of a 'bulky thing' (l. 1) to which Murdoch's speaker first ascribes the 'huge fudge with tentacles upon' (l. 2), only to conjure in the subsequent line the dismal likeness to the interment of a 'dead puppy dog' (l. 3) as she digs the 'stiff clay like a funeral' (l. 5). Murdoch's appalling burial image recalls Christina Rossetti's 'Spring', where life seemingly emerges from the grave too:

> Tips of tender green,
> Leaf, or blade, or sheath;
> Telling of the hidden life
> That breaks forth underneath,
> Life nursed in its grave by Death. (1970, 25)[17]

And, indeed, T.S. Eliot:

> That corpse you planted last year in your garden,

---

[16] There are earlier drafts of 'Gunnera' (PN 8 24–27, 50, 92, 94, 140, 142), and a subsequent draft (PN10 76); 'Gunnera' was first published in *Poetry London / Apple Magazine* 1.1, Autumn 1979, alongside 'Fox' p. 40, 'No Smell', and 'Edible Fungi', pp. 41–2; and the four were republished in *Poems by Iris Murdoch* (*PIM* 88, 87, 86, 89). All four of these poems are also found collected together in a notebook of fair copies (PN13 19–22).

[17] In another Murdoch poem, 'The Brown Horse', Hullah remarks on some 'simple language that weaves a wistful tone redolent of Christina Rossetti's sombre lyrics' (*PIM* 38).

Has it begun to sprout? Will it bloom this year?
Or has the sudden frost disturbed its bed'. (1972, 21–49, l. 71–72)

Jeremy Diaper remarks on the 'repellent undertones' of Eliot's burial image which 'ensures the language of horticulture is conjoined with cadaverous and deathly connotations' (2022, 141). Eliot's horticultural imagery, combining thoughts of both death and regeneration, alludes to a particular societal moment. Written in 1922 following closely after the end of the First World War and the influenza pandemic, 'The Waste Land' can be read as richly symbolic of memory and the transference of that memory through the generations. Murdoch's 'Gunnera' adopts the disquieting image of interment but, unlike Eliot, focuses more directly and more literally on the transformation of the 'bulky thing' (l. 1) she beds in, in winter. With the changing of the seasons, she notes nature's 'Defeat of gravitation / By genial mix of seed and sod' (l. 11–12). In half a year the 'weighty blob' (l. 10), she marvels, 'Has lifted up fantastical in May / A vast rain forest dome' (l. 7–8). Murdoch, in awe of the majesty of the plant, cannot rule out altogether the workings of some sort of ineffable power behind its transformation. But, at the same time, her poem playfully suggests that if its striking 'formal demonstration' is after all part of some ethereal masterplan, then surely God has a 'green grotesque imagination' (l. 13–15). The bog-loving *Gunnera manicata*, native to parts of South America and capable of growing to a height and spread of over two and a half metres, provokes the speaker's wonderment at the richness and diversity of plant life and its potential for rebirth.

If 'Gunnera' invokes the mystery of God and 'His green grotesque imagination', another of the four poems published in 1979 and republished in *Poems by Iris Murdoch*, 'Edible Fungi', contrastively suggests 'primeval forces' and 'pagan exultation', notes Hullah (*PIM* 40).

Edible Fungi
In a muffled wood in a mauve gauze haze
The ground random and dumpy with the cold winter days
Frilled up and frozen in brittled seas of ooze
Our dull footsteps

> Crackle the plates of ice in muddy jars 5
> The print of horses' hooves.
>
> In groups exotic and fastidious
> Fungus has posed its small household
> Tawnyish pink and white
> Graceful as sudden girls, fragile as shells 10
> Or wings its dim gills
> Gratuitously bodied out of night
> Instant and frail this alien flesh
> Ephemeral, from underneath the frost
> And wet womb of the cold. 15
> Now from the pan slimy as fish
> Slithering darkly in their own ink
> We eat water and earth.
> The clean taste of blackness salutes the tongue.
> (PN10 86)[18]

Hullah sees nature presented in 'Edible Fungi' as some sort of 'controller', the poem's 'speaker at best a passive witness', although he also recognizes that the 'natural realm reasserting itself as spring comes to offer "fantastical" rebirth of natural energies' (*PIM* 40). Murdoch may seek to invoke nature's vast and mysterious powers of growth and regeneration in these poems but, rather than regarding this as a form of control over human counterparts, I would suggest that the poem's ecocentric focus prioritizes the powering up of new growth and regeneration occurring independently of, and quite separate from, any human activity: the fungi appear 'Gratuitously' out of the woodland's 'brittle seas of ooze' (l.12, l. 3). 'Gratuitous' and 'ephemeral' mark the independent and contingent activities of the fungus. Once foraged, the inky, earthy 'clean taste of blackness salutes the tongue' (l. 19); the poem ends on a celebratory note that acknowledges the spontaneous bounty. 'We eat water and earth' (l. 18) speaks first to the earthy flavour of fungi, but the final line of Murdoch's 'Edible Fungi' also reiterates the regenerative process: water and soil are the life-giving elements. But it does something more than that. Like

---

[18] There is an earlier draft of 'Edible Fungi' (PN8 52).

Coleridge's 'one Life' and Wordsworth's '*active* principle [...] in all things', the apparently naïve, matter of fact and somewhat repellent line, 'We eat water and earth' (l. 18), acknowledges the complex connections of living things: that we are more involved with other organisms than we ever stop to contemplate, and that the co-reliance of all living things is fundamental to our survival.

'You take life tiptoe. Too swift. Your splendid feet' is the first line of a Murdoch poem that neatly brings together most of the major themes that I have discussed so far. In 'You take life tiptoe', her speaker expresses an embodied sensory interaction with the more-than-human environment, the fleeting nature of life and planetary indifference, and the miraculous regeneration of the natural world devoid of human interference.

[Untitled]
You take life tiptoe. Too swift. Your splendid feet
Hardly touch the year's spinning surface,
Spurning the buoyant earth. Know
Under each toe-print shafting to earth's centre
Existence lies rich—a million fathoms   5
Of shifting splendour song-colour deep. Remember
The blindworm's sensitive cell, the hungry
Sinuous curve of the tree-root
And last year's leaves a-rot—the crushed
Diamond in the making—the curled   10
Heads of flowers underground. These
Spring in you upward,
Blossom and speak in you. Tarry,
And feel the sap-strength a-flow,
Up, up to your hands, to your lips. Cry   15
In salute of life—not in dread
Of dizzy cross-sections of being—relating
All things to all. The black soil
My fingers divide is death,
Yet it crumbles to life in the seed.   20
Horror is real;—but real too
The unashamed certainties—knowledge
Of intricacies of events,

Of the past in the present fixing
The future's invincible waking-point.   25
Strike to the bone and create
All time in a second—for now
You tread tiptoe on tombs and treasure.
1/11/39
(PN4 144)[19]

The speaker demands of her reader to 'Know' (l. 3), 'Remember' (l. 6), 'Tarry' (l. 13), 'Cry' (l. 15), to pause and attend, not only to the planet's natural marvels but to the intricate particularity of what lies beneath our human 'toe-print' (l. 4). The recurring imperatives·placed at line breaks, and scattered irregularly through the first half of the poem, give urgent and rhythmic weight to her insistence that we take account of the more-than-human life that 'lies rich' (l. 5), but out of view, in Earth's 'shifting splendour deep' (l. 6). In consecutive lines, Murdoch effortlessly connects two temporalities—the previous year's decay on the Earth's surface to the vertiginously deep geological time of the billion years that diamonds take to form one hundred miles below the surface of Earth's upper mantle. She pictures sap flowing through the vegetal into the human vein as she exhorts her reader to give time and attention to the emergence of new life springing up from beneath the ground: 'Tarry, / And feel the sap-strength a-flow, / Up, up to your hands, to your lips' (l. 13–15). Before ingeminating the miracle of rebirth and regeneration, Murdoch strikes a positive note, demanding we 'Cry / In salute of life—not in dread / Of dizzy cross-sections of being—relating / All things to all' (l. 15–18). Murdoch's immersive interpretation of new and co-reliant growth in this poem coheres with ecologist Tom Oliver's assertion that humans are 'much more intertwined than we know' with the ecosystem. Using oxygen molecules as his exemplar, he tells us:

---

[19] Iris Murdoch's untitled poem ['You take life tiptoe. Too swift. Your splendid feet'] was first published in the *Cherwell* 57.4 (11 November 1939), and republished as 'Untitled (II)' (*PIM* 63–64). Murdoch composed the poem on 1 November 1939, ten days before it first appeared in the *Cherwell*.

You have incorporated [these oxygen molecules] into you with every breath [...], with the food you have eaten and with the water you have consumed. The molecules have come from farms, rivers and springs from across the entire world. (2020, 21)

These oxygen molecules may have been circulating for aeons in the atmosphere and through various animal bodies, and this compels Oliver to state a surprisingly mundane reality: 'we are comprised of the same commonplace elements that make up everything else in our universe' (2020, 20). Far from being discrete unchanging individuals possessed of immutable traits and acting autonomously in the world, and to echo Bryson's concern regarding 'hyperrationality' (2002, 6–7), we are in fact deeply immersed and intertwined with our environment. Murdoch's 'You take life tiptoe' poem apprehends this integral co-dependence of the living world.

It is interesting, too, to consider this poem alongside a contemporary one simply because of the synergy revealed in their respective endorsements of the human dependence on ecosphere. In this, her third of 'four apocalypse sonnets', Samantha Walton expresses a simple desire in wanting to extend protection even to trees that are dead:

'four apocalypse sonnets' by Samantha Walton
3.
it used to feel so histrionic to invoke the 'apocalypse'
to wield an optimism
as cruel as it is anachronistic
I will take the trees
protecting their dead   5
the slow rotation of the earth
I would give the net of my hair
the queer twirled surface of my spleen
my blood's skein
to be plugged into whatever   10
pulses in the dirt
I will suck the marrow of myself
want only what is best
(2021, np)[20]

---

[20] Reproduced by kind permission of Samantha Walton.

We sense humble desperation when she would give even her 'blood's skein / to be plugged into whatever / pulses in the dirt' (l. 9–11). Both Murdoch and Walton invite their readers to contemplate the incalculable value to all life prompted in Murdoch's case by vegetal 'tombs and treasure' (PN4 114, l. 28), and in Walton's by 'whatever / pulses in the dirt' (2021, l. 10–11). The two poems combine conceptions of the human and the vegetal, suggesting the cycle of life and Earth's power as a self-restoring organism. However, the optimism induced in this reader by the Murdoch poem quickly fades when confronted by the impending and seemingly insurmountable crisis conveyed in Walton's work. Walton's twenty-first-century sonnet is presented here in stark contrast to the mid-twentieth-century context of the Murdochian poem: the picture of environmental anguish as one moves from the earlier poem to the later one can inflict an anxious disquiet.

'The Minute and Absolutely Random Detail of the World'

Nature's alterity complicates any understanding of what it means to be human. Moreover, any experience of the natural world has the capacity to render its perceiver awe-struck; nature's otherness can transcend human consciousness in unexpected moments. Murdoch allows the wildness of the natural world, this otherness, to permeate consciousness and restore a sense of perspective. A 'hovering kestrel' represents, for Murdoch, 'a perfection of form which invites unpossessive contemplation' (SGC 369–70). The kestrel's specific reality, central to her philosophical conception of ethical attention, is rarely discussed in the context of Murdoch's respectful recognition of this raptor's uniquely alien, wild and independent majesty.[21] Environmental philosopher Robin Attfield acknowledges the significance of wild nature's independence (even though he appears not to realize the full force of his analogy when 'windhover'—seventeenth-century British dialect—is synonymous with 'kestrel'):

---

[21] One exception is Silvia Caprioglio Panizza with her talk 'Who Cares about the Kestrel?' at the Wartime Quartet Conference at Durham University, 7–9 June 2023.

Not being planned or devised by humanity, wild nature can remind us of the pettiness of our parochial concerns, or restore our sense of proportion, when we perceive the very indifference of its processes to our own machinations, or when we catch sight of the alien nature of a wild, independent natural creature such as a kestrel as depicted by Iris Murdoch, or such as a falcon, lauded in 'The Windhover' by Gerard Manley Hopkins. (Attfield 2015, 18)

As Attfield points out, 'awareness of nature's otherness is premised on the natural environment being there to be discovered, transcending our consciousness, and by no means our creation' (2015, 18). Hopkins and Murdoch marvelled at the natural world: while Hopkins celebrated its divine creation and had faith in divine protection, Murdoch's was a more grounded appreciation of earthly beauty and nature's potential as a site for moral change.

Murdoch was a great admirer of her Victorian predecessor despite their theistic differences, notably, their contrastive convictions of a transcendent God in Hopkins's case, and in Murdoch's secular belief in an immanent Good. Murdoch's own collection of Hopkins's poetry in the Iris Murdoch Collections (IML 191) is not annotated, as Paul Fiddes rightly remarks (2022, 82), but she has recorded a general comment about her sense of deep connection with Hopkins. A journal entry attempts to compare literature's different 'kinds of pleasure' (J3 29). She is making an affective distinction between the 'delight' found in a writer's work where 'one [remains] undivided' or unmoved, and the moment when 'one is "touched"' and 'one [becomes] divided' (J3 29). She places Hopkins decisively in the second category, explaining that here 'it is different, one is touched, it is related to oneself. One thinks about it? It inspires—humility. [...] O[ne] has a dialogue with oneself, o[ne] sees oneself' (J3 29). One can surmise from this entry that Hopkins's approach to poetry represented a significant source of inspiration for Murdoch. Hullah notes an 'Hopkinsian opening' (*PIM* 34) to her 'Poem', first published in the *Cherwell* on 13 May 1939, that begins, 'Lovely is earth now, splendid / With year-youth' (PN4 107; *PIM* 60). Here, she seems to be summoning Hopkins's celebratory poem 'Spring' (1877): 'Nothing is so beautiful as spring', producing its 'strain of the earth's sweet being in the beginning' in the second verse (*GMH* 28). Now that Murdoch's poetry notebooks are available for reading in the Iris Murdoch Collections, observant researchers will doubtless identify further sources of poetic inspiration in this work.

During her lifetime, Murdoch's poetry was to remain a largely private endeavour, despite the encouragement of others, including John Bayley. When Yozo Muroya suggested that Murdoch publish some poems, Hullah remembers that 'she was clearly not entirely against the idea. John Bayley was very enthusiastic (in that way he had of being very enthusiastic over just about anything). A seed had been planted' (Hullah, 2020). Some of her best poems are yet to appear in print. Many of Hopkins's best poems were not published during his lifetime either: his departure from the poetic style of his Victorian contemporaries was considered to be sufficiently radical that it took until 1918 (twenty-nine years after Hopkins's death) for his friend and literary executor, Robert Bridges, to bring much of Hopkins's surviving work to publication (*GMH* xiii–xvi). Hopkins's particular poetic genre and legacy, best categorized as religious nature poetry, comes from the compulsion to combine nature into poetry grounded in his firm belief in a transcendent God. Hopkins's faith makes Murdoch's admiration for him particularly intriguing when, despite venerating many of Christianity's rituals, she felt unable to share his conviction.

On 1 March 1969, thirty years before Attfield connects Murdochian kestrel and Hopkinsian windhover, Murdoch appears baffled by the theological link Father Martin Jarrett-Kerr makes during a BBC discussion: 'He rather touchingly connected my "hovering kestrel" in the Leslie Stephen Lecture to Gerard Manley Hopkins's "Windhover"—which represents the Holy Ghost!', she exclaims in her journal (J10 168). This bafflement stems from the mischaracterization of the diametrically opposed theistic perspectives of the two writers. Transfixed as they both may be by the kestrel, further comparison produces in fact an interesting symbiosis of their separate ideas. In her philosophical essay 'The Sovereignty of Good over Other Concepts', Murdoch invokes the spellbinding sight of a hovering kestrel to convey her picture of attention to the other that forces us out of our doleful self-obsession: 'The brooding self with its hurt vanity has disappeared. There is nothing now but kestrel'

(SGC 369). This image of close attention to nature is echoed in Murdoch's poetry. For Hopkins, for whom the natural world is also sacred, both God and Earth deserve veneration. This synchronic approach to his writing (combining nature and God) produced two of his most well-known poems, 'The Windhover' (1877), which Hopkins dedicates 'To Christ our Lord', and the curtal sonnet 'Pied Beauty' (1877). 'Pied Beauty' is an incantation to the divine beauty found here on Earth:

> Glory be to God for dappled things—
> For skies of couple-colour as a brinded cow;
> For rose-moles all in stipple upon trout that swim. (*GMH* 128)

Hopkins evokes the natural beauty of a setting in which, in the second tercet, he carefully places remnants denoting human impact:

> Fresh-firecoal chestnut-falls; finches' wings
> Landscape plotted and pieced—fold, fallow, and plough;
> And áll trádes, their gear and tackle and trim. (*GMH* 128)

At this juncture, Hopkins appears to view such activity as a form of earthly stewardship and appears to hold firm to his conviction of heavenly protection and renewal, proclaiming, 'He fathers-forth whose beauty is past change: / Praise him' (*GMH* 128). Early criticism of Hopkins's work, his departure from standard sonnet forms (and the subsequent ruminations over the acceptance—or not—of his 'curtal' form), once distracted from the sonnet's content. What we are able to celebrate today is his lyrical acknowledgement of the quiddity and, indeed, the beauty of his natural surroundings.

Murdoch, by contrast, takes a secular and earthly approach to her work which makes her admiration for Hopkins all the more surprising. She aspired to what she terms 'good art' as a means of acknowledging and conveying the living world's particularity. She says, 'Good art reveals what we are usually too selfish and too timid to recognize, the minute and absolutely random detail of the world, and reveals it together with a sense of unity and form' (SGC 371). Murdoch saw art as a means to derive meaning in a contingent real world. By contrast, Hopkins appears to seek out pattern in nature. He believed in the predetermined character of each living thing and in an almost Ruskinesque unity of science and aesthetics. He pursued nature's dynamic patterns. 'I was looking at high waves', Hopkins notes in his journal on 10 August 1872:

> the breakers always are parallel to the coast and shape themselves to it except where the curve is sharp however the wind blows. They are rolled out by the shallowing shore [...]. The slant ruck or crease one sees in them shows the way of the wind. The regularity of the barrels surprised and charmed the eye. (*GMH* 128)

However, Fiddes argues that Hopkins's universe, like Murdoch's, was also 'full of individual forms', too, and the poet celebrated particularity:

> Hopkins [...] not only *observes* the particularity of 'pied beauty' in the world, but also *creates* beauty in an art-object which has its own distinct 'pied' character. In every way he fulfils the requirement of Murdoch that an artist should give sustained attention to the contingent details of the world. (2022, 102–3)

Particularity reveals itself when, for example, each 'stone makes its own kind of splash when thrown into a well; each bell makes its own ring' (Fiddes 2022, 103). To take Fiddes's image further, the pattern for Hopkins would emerge in the confidence that the stone would splash and the bell would ring. Hopkins called this patterning 'inscape', and inscapes were powered by stress or 'instress', a God-given energy received by the observer that became the instrumental factor in establishing inscape (Siddall 2009, 71).

In his book *Green Man Hopkins: Poetry and the Victorian Ecological Imagination*, which offers a comprehensive assessment of the poet and his significance to Victorian ecology, John Parham explains that these two concepts, 'inscape' and 'instress', devised during Hopkins's undergraduate years, represent 'the creation, in advance of his poetry, of an aesthetic paradigm analogous to the two components of ecosystem theory—physical being and the energy that holds it together' (2010, 53). Parham says that the poet's 'interest in nature was informed by a contemporary science that was gradually shifting from more static models—natural history, botany—to an understanding of nature as an energy system characterised by movement and flux' (2010, 53). Hopkins's intuitive sense of 'instress', in the idea of charged elements of landscape and of objects responding in the world, calls to mind the animistic sensibility of Murdoch's later fiction that I discuss in the next chapter. Hopkins's belief in a God-given design to the natural world, in circular fashion, affords him the courage of his own conviction. While Murdoch accepted, even celebrated, Earth's mysterious and often unfathomable quiddities, Hopkins was on a mission to identify nature's laws and to connect divine power in the determination of those laws. For Hopkins, then, faith in God is the great unifying force. Such a notion makes his deep disappointment in his fellow man all the more poignant in 'God's Grandeur' (1877): 'Why do men then now not reck his rod?' (*GMH* 27). Why not take heed of the grandeur and beneficence of God? As far as Hopkins is concerned, humankind seems intent on overlaying and concealing God's work:

> And all is seared with trade; bleared, smeared with toil;
> And wears man's smudge and shares man's smell: the soil
> Is bare now, nor can foot feel, being shod. (*GMH* 27)

Despite the burgeoning industrialization of the Victorian age, Hopkins's firm belief is that nature can prevail under the protection of God's grandeur: 'And for all this, nature is never spent; / There lives the dearest freshness deep down things'; and will 'flame out' (*GMH* 27). (Mis)quoting this line in her journal, Murdoch recalls 'God's Grandeur': '"There lives a deeper freshness deep down things" G.M.H.' (J14 23). It is a hopeful and uncharacteristically upbeat entry that she records on 12 August 1982, on returning home from a holiday in France and recalling a trip to the Camargue to see the flamingos. Murdoch may not share Hopkins's conviction of an Absolute Being, but she seems drawn nonetheless to the clarity of his Christian faith and, of course, to his profound love of the natural world. Hopkins derived a certain peace in writing at the intersection of nature and Christian belief. Later on, however, there would be no such harmony between the renewed commitment he made to God when he converted to Catholicism and eventually ordained a Jesuit priest, and his restlessly creative poetic soul which is articulated in 'My heart is hiding / Stirred for a bird' (*GMH* 30). Stirred to serve his God, the poet submitted to his new ecclesiastical calling, eventually leaving poetry behind.

The hovering kestrel is an image that stays with Murdoch. In *Henry and Cato* (1976), her eighteenth novel, a kestrel is to be seen hovering over wasteland on three occasions. Then, two years later, a kestrel appears in a poem published in the calendrical collection *A Year of Birds* (1978), a collaboration between Murdoch, the poet, and her friend, the engraver, typographer and artist Reynolds Stone. *Henry and Cato* concerns two childhood friends, art historian Henry Marshalson and Cato Forbes, a Catholic convert and priest who, as the story begins, is struggling with his faith and with his love for Beautiful Joe. Henry and Cato have their own reasons to be introspective and caught up in their own fantasies. Both men are preoccupied; neither pays attention. The fleeting and reprised appearance of the kestrel in the novel evokes, for some readers of Murdoch at least, the 'hovering kestrel' episode (SGC 369–70) in the philosophy about attending to the other, having the vision to see clearly, so long as self-obsession or self-delusion does not get in the way. In the novel, Henry catches sight of a kestrel twice, and then Cato sees one moments after burying his cassock:

> The kestrel was perfectly still, an image of contemplation, the warm blue afternoon spread out behind it, vibrating colour and light. Cato looked at it, aware suddenly of nothing else. Then as he looked, holding his breath, the bird swooped. It came down, with almost slow casual ease, to the ground, then rose again and flew away over Cato's head. […] 'My Lord and my God', said Cato aloud. Then he laughed and set off again in the direction of the Mission. (*HC* 186)

The kestrel heralds Cato's transition from the 'searing sense of loss and shame' of his erstwhile faith, to 'feeling the power of love' (*HC* 188) as he dares in that moment to contemplate the possibility of a future with Beautiful Joe.

In an early draft of a poem composed in 1972, ['Over the wispy yellow slopes of the motorway'] that eventually comes to publication as

'August' in *A Year of Birds*, Murdoch's speaker conjures the image of a motorist, travelling along a motorway lined (most likely) by fields of golden rape, who suddenly catches sight of a gloriously ethereal hovering kestrel. In this early draft, Murdoch conceives of the potential danger in the proximity of the 'motionless' raptor (PN7 103, l. 4) to the fast road.

> **An untitled early draft of the August verse**
> Over the wispy yellow slopes of the motorway
> The August traveller released to holiday
> Sees suddenly air-perched & perilous
> Its motionless meditation aloof from the terrible tarmac
> The oh so frail eternal hovering kestrel   5
> (PN7 103)

In her picture of the motorist's sudden attention to the hovering creature, she conveys a sense of the kestrel's frailty but, by engaging the descriptor 'eternal' (l. 5), she acknowledges its everlasting majesty and its power to command attention too. In the published version of the poem 'August', the motorway is no longer presented as a corridor cleaved through golden rape.

> **August,** *A Year of Birds*
> By the bleached shoulder of the motorway
> The August traveller released to holiday
> Sees suddenly the portent perched in air,
> In meditation aloof near the lethal tarmac
> The moveless flutter of the fragile kestrel.   5
> (*YB* np)

Instead, 'By the bleached shoulder of the motorway' (l. 1) conjures a somewhat more hostile picture of the perilous highway. The already 'terrible tarmac' of the early draft has now become 'lethal' (l. 4). The traveller, again 'released to holiday' (l. 2), encounters the raptor aloft but the picture is somewhat less joyful in the verse's final iteration. The speaker ascribes a sense of foreboding to the skyward object of attention: the 'hovering kestrel' of the original draft has become, at once, 'moveless' and 'fragile' (l. 5). In sum, the published quintain carries portentous undertones of threat to life inferred by the motorway's 'lethal tarmac' for both traveller and kestrel, attributed to human enterprise (l. 4); most poignantly of all, the speaker no longer ascribes eternity to this consistently attention-diverting spectacle.

Despite the differing visions of Hopkinsian windhover and Murdochian kestrel, both demand moral attention: 'shéer plód' (*GMH* 30). Murdoch puts her heart on show in an emotional response to her kestrel with 'self-forgetful pleasure' (SGC 369), but pictures attention to the other as continuous and unending work in one's 'ability to sustain clear vision' (SGC 373). Hopkins's faith in God remains unwavering despite witnessing the threat to nature in the 'felled, felled, are all felled' trees in 'Binsey Poplars' (1879), that 'Quelled or quenched in leaves the leaping sun'; the lines 'O if we but knew what we do / When we delve or hew— /Hack and rack the growing green!' and 'After-comers cannot guess the beauty been' voice his despair at the trees' irrevocable loss (*GMH* 39–40). In 'Inversnaid' (1881), in which he celebrates the Scottish Highland landscape two years later, he asks, 'What would the world be once bereft / Of wet and wildness?'; instead, he urges, 'Let them be left, wildness and wet; / Long live the weeds and the wilderness yet' (*GMH* 50–51). These entreaties to preserve nature's wildness become heightened concern in 'The Sea and The Skylark' (1877) where Hopkins points to the environmental degradation brought about by the creeping urbanization of the English landscape.

In 'The Sea and The Skylark', Hopkins addresses the ironic dissonance of the timeless 'noises' of the sea including the skylark's call ('too old to end') which have to compete with the incoming din of human progress. 'How these two shame this shallow and frail town!'; the 'pure' sounds of the sea and the skylark no longer fit a 'sordid turbid time' where human construction is nature's destruction (*GMH* 29). Decrying the loss of 'earth's past prime' in the wake of mounting industrialization, Hopkins delivers the powerful oxymoron in reduced rhyme—'Our make and making break'—in the poem's final tercet (*GMH* 29). Hopkins's profound faith in God and his belief in the divine intervention in nature's patterning eventually prove out of kilter with the march of progress.

In a similar way, this next Murdoch poem provides a commentary on the hostile forces that work against nature. It comes from the poetry notebook she kept while a student at Somerville College, Oxford. 'The

City in the Plain' describes a visit to a stretch of the River Thames at Oxford meandering through a 'sullen suburb' (l. 2) in the trailing sun of winter; the water is rendered 'malicious' (l. 3) through the careless bespoiling of human detritus: 'oily ripples' (l. 3) and 'broken bottles' (l. 6).

The City in the Plain

Today's sun trails a transient beauty over
The sullen suburb with its bitter trees—
And the oily ripples of the malicious river
Which lacks the savour & the splendour of the seas.
The small wave taps the empty walls forever   5
Bearing its broken bottles & its broken gulls.
It has no song to sing for the lost & the lover—
The city festers in a million cells.

Yet one pure heart would have warded off from the walls
The hailing of hell when the plain shook end to end.   10
Out of the shadows where men sighed and sinned
The green tree grows & grows to greet the showers.
We see when the house cracks & the tree falls
Flora's white feet still moving in the flowers.
2/12/42
(PN5 67)

The bleak dismissal at the end of the first stanza, 'The city festers in a million cells' (l. 8), referring to the anonymous masses of the city's human inhabitants, presents a stark contrast to Murdoch's loving descriptions of non-human subjects, for example, the 'blindworm's sensitive cell' or the 'hungry / sinuous curve of the tree-root' (l. 7–8) of her 'You take life tiptoe' poem. With 'one pure heart' (l. 9) her speaker in 'The City in the Plain' connects a singleness of heart towards God with care for the environment; it suggests that a loss of faith in one produces a loss of care of the other. Despite evident disconsolation at the start of this poem, a joy at regeneration emerges: the second stanza celebrates the persistence of the 'green tree' that 'grows & grows to greet the showers' (l. 12) despite the evident challenges. Even as agèd buildings crumble and trees fall,

'Flora's white feet' (l. 14) continue to regenerate the landscape unrestrained. Her speaker's dismay at humankind's desecration is set in stark contrast to her joy in nature's perceived powers of regeneration.

The Loss of Tender Night

Hopkins's anxiety about human alienation from a bounteous Earth is not present in his early poem 'The Nightingale' (1866). Here, the persistent sound, the 'warbling of the warbling bird', dominates the early dawn:

> For he began at once and shook
> My head to hear. He might have strung
> A row of ripples in the brook,
> So forcibly he sung. (Phillips 2002, 79–80)

Almost a century later and reminiscent of Hopkins's objection to human progress masking the call of the skylark, one cannot fail to observe in a poem by Murdoch how the nightingale's distinctive birdsong is drowned out by the urban clamour of a busy road.

> **Rendezvous with nightingales, for Stella 24 May 1955**
> Stopping the car, we heard the nightingale.
> And climbed the ploughed field from the busy road.
> The night was dark & blue & waiting there.
> Below us an illuminated trail
> Of cars & lorries east & westward flowed   5
> While we were left alone to mount the stair
> Of earth toward the voice upon the ridge
> Beyond the pallid blackthorn of the hedge.
>
> We reached to top with clay upon our shoes
> Drawn by that pure serrated line of sound.   10
> Before us lay the structure of the wood
> First deep in grass & then upon the moss
> Then upon beech leaves seeking a way round
> We skirted fences till at last we stood

Close to that cry so piercing & so lovely 15
Within a cavern made of darkness only.

18 December 1958
But that was long ago; and all the joy
Each had of each that led us to that place
Is dead, & we to other loves do turn. 20
In just such pictures you I still enjoy,
Revisiting your dark & spellbound face
And in your heart in equal image burn.
When detailed memory of our passion fails
We meet each other with the nightingales. 25
(PN6 136)

'Rendezvous with Nightingales' identifies the urban encroachment on the natural world as a concern in the 1950s. Half a century on, in the mid-2000s, Macfarlane calculated that only a 'small and diminishing proportion of terrain [was] now more than five miles from a motorable surface', noting that burgeoning urbanization was becoming an increasing problem for wildlife in Britain (2007, 9). It seems impossible today to conceive of the natural unsullied calls and signals of wildlife that Hopkins, and to a lesser extent Murdoch, would have had the opportunity to experience.

In 'Rendezvous with Nightingales', Murdoch and her companion appear to know where they might encounter nightingales, but not because they can hear their distinctive call. Only when the car is parked up and they have walked away from the 'busy road' (l. 2) do they have any sort of chance of an encounter. The speaker and her companion climb upward, leaving behind the noise and light pollution of the 'illuminated trail / Of cars and lorries' that 'east and westward flowed' (l. 4–5). The enduring richness of the night ('The night was dark & blue & waiting there' [l. 3]) has been, until this moment, obscured by the dazzling flood of artificial light. The poem's narrative progression, the upward climb of poet and companion embedded in iambic pentameter, compels the two women and her reader towards the nightingales. Given the urbanized soundscape, only once away from the road can the pair be 'Drawn by that pure serrated line of sound' (l. 10). It is no coincidence that the iambic rhythm falters here: 'Close to that cry so piercing & so lovely / Within a cavern made of darkness only' (l. 15–16). The effect of this faltering rhythm is to highlight the now dominant and pure lilting sound of intermittent trills, gurgles and whistles with which the nightingale is able to captivate its audience. Murdoch joins a long tradition in attending to the nightingale's call.

The nightingale's seductive sounds penetrating the darkness recalls John Keats's famous meditation: the beautiful sound provoking a web of thoughts from his speaker on beauty, nature, time and death. The 'inviolable voice' of Eliot's nightingale invokes, via Milton's 'sylvan scene', the myth of Philomel already immortalized by Ovid, and heralds 'The change of Philomel, by the barbarous king/ So rudely forced' in 'The Waste Land' (1972, 21–49).[22] In 'Rendezvous with Nightingales' the pained beckoning of the sharp 'serrated line' (l. 10) is resisted by a second line of exclosure, 'fences' that must be 'skirted' (l. 14). What remains of the nightingales' habitat proves almost impenetrable.

The first iteration of Murdoch's poem ends at the second octave. The speaker, imagining herself in a Platonic cavern 'made of darkness only' (l. 16) and removed from the highway, attempts to conjure the sort of sublime sense of purity, recalled in Keats's 'tender is the night' (2007, 193–95, l. 35), that enhances the senses and summons the truth of the experience. The metrical arrangement of 'Rendezvous with Nightingales' evokes the sonnet form. The tender encounter of the two women is a love song to an intense shared moment of hard-won attention to a creature in danger of being obscured and overlooked by human activity. The choice of 'Rendezvous' in the poem's title not only refers to these encounters, but carries a more all-encompassing interpretation that includes intention: a meeting that requires planning, the nightingales sought out. 'Rendezvous' can also express the choice of a specific location and may suggest an element of secrecy surrounding the assignation. 'Rendezvous with Nightingales', written on 24 May 1955, in fact recounts a real event and the first two octaves are dedicated to 'Stella', with whom Murdoch shared this experience. Peter J. Conradi relates how, on 'a drive from London to

---

[22] Eliot's fragmentary modernist poem draws on Milton (2003, l. 140), and Ovid's 'Tereus and Philomela' (2010, 157).

Oxford up the old A40, [Stella] Aldwinckle told Iris to stop the car in a layby; they walked half a mile to a wood where nightingales were magically singing' (*IMAL* 307–8); Conradi does not share the source of his account, or make a connection to the poem.

The third and final octave added to the poem by Murdoch three and a half years later provides the volta, the rhetorical shift, in 'Rendezvous'. Given the passing of time, the last stanza inevitably proves reflective. The relationship between the women, alluded to in the poem's first two stanzas with the intense and exclusive use of pronouns 'we', 'us' and 'our', is confirmed to be long over, 'all the joy / Each had of each that led us to that place / Is dead' (l. 17–19). The visceral recollection stays with the speaker, as it concludes with the couplet 'When detailed memory of our passion fails / We meet each other with the nightingales' (l. 24–25). The women's relationship is memorialized in the beauty of this shared encounter.

In Murdoch's sixteenth novel, *The Sacred and Profane Love Machine* (1974), urban encroachment is proceeding apace: the M40 motorway is under construction along the same stretch of road between Oxford and London where Murdoch and Aldwinckle once found nightingales. Blaise and Harriet Gavender's son, David, has learned that his father's lovechild, Luca, is to be given a bedroom at Hood House. David's sense of intrusion and associated disruption is mirrored by the coming of the motorway to their peaceful Buckinghamshire valley:

> a few lorries and figures of men marked the further progress of the monster, soon now to link up with another monster and become a humming lifeline of the New Britain, banishing silence forever, day and night, from that peaceful valley. These were the last days of the silence. (*SPLM* 174)

David strolls onto the half-constructed carriageway and lies on his back with 'the sun dazzling into his eyes, a huge blue quiet heaven above him traced over by white trails of high soundless aeroplanes' and he worries: 'How irrevocably spoilt, down to its minutest detail, his world was now. Even the countryside was spoilt, the animals, the birds, the flowers. There was nowhere to run to. Poor innocent toads had been desecrated forever' (*SPLM* 174). Later in the novel, David retraces his steps to the motorway, his disappearance intended to discourage his mother from leaving for the airport:

> The concreted courtyard where he had once lain supine in the sun in a final act of solitude was now a racing track of glittering motor-cars. And upon the nearer carriageway as he approached he could see, and shuddered at it, the squashed and flattened form of a hare, a monogram of fur and blood. (*SPLM* 282)

The scene presages a rather grim ending to the novel, but it and David's earlier visit to the motorway are included here to demonstrate the small but significant elements of thematic alignment with Murdoch's early poem 'Rendezvous with Nightingales'.

It has been argued that Murdoch's poetry was never intended for publication, but instead a ritual that enabled her to explore the inner life; of the poems written for Wallace Robson, for example, White avers that they were 'to help Murdoch discover her true feelings', and connect with the world around her (*IMR*5 8). As Wordsworth's contemporary William Hazlitt once observed, poetry is the 'universal language which the heart holds with nature and itself' (1892, 9–10). However, one can feel less confident of Murdoch's true intentions when one considers how she had already published many poems as a young woman, others she kept revising and, of course, towards the end of her life she collaborated with Hullah and Muroya over *Poems by Iris Murdoch*. Nonetheless, when one stops to reflect once more on that final contemplative octave that Murdoch added to 'Rendezvous' later on, it certainly stands to remind this reader, at least, of the deeply personal nature of some of her poetry.

For Murdoch, composing poetry was 'an almost impossible art' (Bigsby 1995). She was aware of what she saw as her limitations as a poet, and she was modest about her achievements. Martin and Rowe record that by the 'end of the 1980s she had called her experiments with poetry to a halt' (2010, 135); she asked Chatto & Windus to return the '100-page cycle of poems, "Conversations with a Prince"', that she had submitted to them in the 1960s (2010, 71). Yet, on the subject of her poetry, Rowe avers that one ought never to overlook those 'brief moments of magical awareness of the world that Murdoch is able to conjure with a clarity of perception' (*IMN*13 15). This chapter has scrutinized a few (of the many) poems by Murdoch that offer themselves up to ecopoetic interpretation. The poems have been situated in the context of those who inspired her, and with whom her work aligns thematically. The great significance of Murdoch's nature poetry is in providing an important and insightful prequel to what are to become key ecological markers in the fiction-writing that follows: increasingly animistic landscapes, the spaces she creates for the independent lives of other creatures and, most significant of all, the depiction of solitary moments of epiphany for key characters in landscape.

References

Primary Sources: Featured Poems by Iris Murdoch from the Poetry Notebooks (PN), KUAS202/3/4–13 in the Iris Murdoch Collections

Murdoch, Iris, untitled poem ['The morning fills my eyes & my heart'], 4 February 1938a, PN4 12

———, untitled poem '['What does it matter'], 8 February 1938b, PN4 14

———, untitled poem ['Brilliant oaktrees, gold mixed with their green'], 1 July 1941, PN5 43

———, 'The City in the Plain', 2 December 1942, PN5 67

———, untitled poem, ['Gently I have touched the thin lids of your eyes'], 23 August 1943, PN5 226

———, 'Rendezvous with Nightingales', 24 May 1955 and 18 December 1958, PN6 136

———, untitled poem 'for WR', ['The trailing stars tell of dooms'], PN6 12

———, untitled poem ['Snowdrops smell'], PN8 13

———, 'Gunnera', PN8 148

———, 'Edible Fungi', PN10 86

———, untitled poem, ['You take life tiptoe. Too swift. Your splendid feet'], PN4 144

———, untitled poem, ['Over the wispy yellow slopes of the motorway'], PN7 103

Primary Sources: Novels

Murdoch, Iris. 1999. *The Sea, The Sea* (*TSTS*) (1978). London: Vintage
———. 2000. *The Message to the Planet* (*MP*) (1989). London: Vintage
———. 2002. *A Word Child* (*WC*) (1975). London: Vintage
———. 2002. *Henry and Cato* (*HC*) (1976). London: Vintage
———. 2003. *The Sacred and Profane Love Machine* (*SPLM*) (1974). London: Vintage

Primary Sources: Philosophical Writings

Murdoch, Iris. 1997. *Existentialists and Mystics: Writings on Philosophy and Literature* (*EM*), ed. Peter J. Conradi. Harmondsworth: Penguin
———. 1997. 'Literature and Philosophy: A Conversation with Bryan Magee' (LP). In *Iris Murdoch, Existentialists and Mystics: Writings on Philosophy and Literature*, ed. Peter J. Conradi, 3–30. London: Penguin.
——— 1997. 'On "God" and "Good"', (OGG), In Iris Murdoch, *Existentialists and Mystics: Writings on Philosophy and Literature*, ed. by Peter J. Conradi, 337–362. London: Penguin.
——— 1997. 'The Sovereignty of Good over other Concepts' (SGC), In Iris Murdoch, *Existentialists and Mystics: Writings on Philosophy and Literature*, ed. by Peter J. Conradi. 363–385. London: Penguin.

Primary Sources: Previously Published Poetry

Murdoch, Iris. 1952. Untitled poem for WR ['The trailing stars tell of dooms'] (PN6 12) first published in Frances White (2014) 'Raids on the Inarticulate: Poems for Wallace Robson'. *Iris Murdoch Review* 5, 8–16
Murdoch, Iris, and Reynolds Stone. 1984. *A Year of Birds* (1978). London: Chatto & Windus
Muroya, Yozo, and Paul Hullah, eds. 1997. *Poems by Iris Murdoch* (*PIM*). Okayama: University Education Press.

Primary Sources: Other Work

Murdoch, Iris. 1970. A Note on Drama. *Cue: Greenwich Theatre Magazine* (September 1970), 13–14, KUAS139/2, from the Iris Murdoch Collections.

Primary Sources: Journals from the Iris Murdoch Collections

Murdoch, Iris. Journal 3 (J3), 4 June 1945–12 May 1947, KUAS202/1/3
———. Journal 9 (J9), 30 March 1954–24 February 1964, KUAS202/1/9
———, Journal 10 (J10), February 1964–18 March 1970b, KUAS202/1/10
———. Journal 11 (J11), 18 March 1970–1916 May 1972, KUAS202/1/11
———, Journal 12 (J12), 1 April 1975–23 May 1978, KUAS202/1/12
———, Journal 14 (J14), 1 January 1981–8 August 1992, KUAS202/1/14

Secondary Sources: Books, Chapters, Journal Essays, Newspapers, Poetry Collections, Sound Recordings, Websites

Attfield, Robin. 2015. *The Ethics of the Global Environment*. 2nd ed. Edinburgh: Edinburgh University Press.
Auden, W. H. 1951. *The Enchafèd Flood or The Romantic Iconography of the Sea*. London: Faber and Faber, IML 81, in the Iris Murdoch Collections.
———. 1979. *W. H. Auden: Selected Poems*, ed. Edward Mendelson. London: Faber and Faber.
Badminton School. 1938. *Badminton School* magazine No. 77, Summer Term 1938, KUAS6/10/3
Bate, Jonathan. 1991. *Romantic Ecology: Wordsworth and the Environmental Tradition*. London: Routledge.
Bavidge, Jenny. 2019. A Sense of Climate Crisis Now Haunts Stories Which Aren't Even about the Environment. *The Conversation*, 24 September 2019. https://theconversation.com/a-sense-of-climate-crisis-now-haunts-stories-which-arent-even-about-the-environment-123429. Accessed 24 September 2019.
Bigsby, Christopher. 1995. *Kaleidoscope*, BBC Radio 4
———. 2003. Interview with Iris Murdoch. In *From a Tiny Corner in the House of Fiction: Conversations with Iris Murdoch (TCHF)*, ed. Gillian Dooley, 97–119. Columbia SC: University of South Carolina Press.
Blake, William. 1971a. XXVIII *The Sick Rose*. Poems from the Notebook (c. 1791–2) in W.H. Stevenson, ed. *The Poems of William Blake*. London, Longman.

———. 1971b. 'Book of Thel' (c. 1789–1790). In *The Poems of William Blake*, ed. W. H. Stevenson. London: Longman.

Bradford, Richard. 2010. *Poetry: The Ultimate Guide*. Basingstoke: Palgrave Macmillan.

Bridges, Robert, ed. 1930. *Poems of Gerard Manley Hopkins*, 2nd ed. London: Oxford University Press, IML 191, in the Iris Murdoch Collections

Bristol School Boys and Girls. *Poet Venturers: A Collection of Poems*, KUAS79, in the Iris Murdoch Collections

Bryson, J. Scott, ed. 2002. *Ecopoetry: A Critical Introduction*. Salt Lake City: University of Utah.

Coleridge, Samuel Taylor. 1997. *The Complete Poems*, ed. William Keach. London: Penguin

Commoner, Barry. 1971. *The Closing Circle: Nature, Man and Technology*. New York: Knopf.

Conradi, Peter J. 2001. *Iris Murdoch: A Life (IMAL)*. London: HarperCollins.

Diaper, Jeremy. 2022. Planting, Gardens and Organicism. In *Eco-Modernism: Ecology, Environment, and Nature in Literary Modernism*, ed. Jeremy Diaper, 139–151. Clemson SC: Clemson University Press.

Eliot, T. S. 1972. *The Waste Land and other poems*. London: Faber and Faber.

Felstiner, John. 2009. *Can Poetry Save the Earth? A Field Guide to Nature Poems*. New Haven: Yale University Press.

Fiddes, Paul S. 2022. *Iris Murdoch and the Others: A Writer in Dialogue with Theology*. London: T&T Clark.

Gardner, W. H., ed. 1953. *Gerard Manley Hopkins: A Selection of his Poems and Prose (GMH)*. London: Penguin.

Garrard, Greg. 2012. *Ecocriticism*. 2nd ed. Abingdon: Routledge.

Gifford, Terry. 2011. *Green Voices: Understanding Contemporary Nature Poetry*. 2nd ed. Nottingham Critical: Cultural and Communications Press.

Grene, Marjorie, ed. 1969. *The Anatomy of Knowledge*. London: Routledge & Kegan Paul.

Hazlitt, William. 1892. *Lectures on the English Poets*. New York: Dodd, Mead and Company.

Heptonstall, Geoffrey. 1999. The Poetry of Iris Murdoch. *The Contemporary Review* 1 February 1999; 274, 1597; Periodicals Archive Online, 84–89

Horner, Avril, and Anne Rowe, eds. 2015. *Living on Paper: Letters from Iris Murdoch 1934–1995 (LoP)*. London: Chatto & Windus.

Hullah, Paul. 1997. Critical Introduction. In *Poems by Iris Murdoch (PIM)*, ed. Yozo Muroya and Paul Hullah. Okayama: University Education Press.

———. 2020. Email correspondence, 28 January 2020, reproduced by kind permission of the author.

———. 2023. "Now the Illumination": Iris Murdoch as Zen Philosopher-Poet. In *Iris Murdoch and the Literary Imagination, Iris Murdoch Today*, ed. Miles Leeson and Frances White. London: Palgrave Macmillan.

Hutchings, Kevin. 2007. Ecocriticism in British Romantic Studies. *Literature Compass* 4 (1): 172–202. https://doi.org/10.1111/j.1741-4113.2006.00417.x.

Keats, John. 2007. *Selected Poems*. London: Penguin.

Lidström, Susanna, and Greg Garrard. 2014. "Images Adequate to Our Predicament": Ecology, Environment and Ecopoetics. *Environmental Humanities* 5:35–53. https://doi.org/10.1215/22011919-3615406.

Mac Cumhaill, Clare, and Rachael Wiseman. 2022. *Metaphysical Animals: How Four Women Brought Philosophy Back to Life (MA)*. London: Chatto & Windus.

Macfarlane, Robert. 2007. *The Wild Places*. London: Granta.

Martin, Edward, and Bill Pickard, eds. 1973. *600 Years of Bristol Poetry*. Bristol, Arts and Leisure Committee of the City and County of Bristol, KUAS6/10/2/7, Iris Murdoch Collections

Martin, Priscilla, and Anne Rowe. 2010. *Iris Murdoch: A Literary Life*. Basingstoke: Palgrave Macmillan.

Milton, John. 2003. *Paradise Lost*. London: Penguin.

O'Flinn, Paul. 2001. *How to Study Romantic Poetry*. 2nd ed. Basingstoke: Palgrave.

Oliver, Tom. 2020. *The Self-Delusion*. London: Orion.

Ovid. 2010. *Metamorphoses*, trans. and ed. Charles Martin. London: W.W. Norton & Co.

Parham, John. 2010. *Green Man Hopkins: Poetry and the Victorian Ecological Imagination*. Amsterdam: Rodopi.

Phillips, Catherine, ed. 2002. *Gerard Manley Hopkins: The Major Works*. Oxford: Oxford University Press.

Read, Catherine, and Agnes Szokolszky. 2020. Ecological Psychology and Enactivism: Perceptually-Guided Action vs. Sensation-Based Enaction. *Frontiers in Psychology* 11. https://doi.org/10.3389/fpsyg.2020.01270.

Read, Daniel. 2019. *The Problem of Evil and the Fiction and Philosophy of Iris Murdoch* (Unpublished Thesis, Kingston University, June 2019).

Reilly, Evelyn. 2013. Environmental Dreamscapes and Ecopoetic Grief. In *OmniVerse*. http://omniverse.us/evelyn-reilly-environmental-dreamscapes-and-ecopoetic-grief/. Accessed 19 August 2021.

Rossetti, Christina. 1970. Spring. In *A Choice of Christina Rossetti's Verse*, ed. Elizabeth Jennings. London: Faber and Faber.

Rowe, Anne. 1999. Review of Poems by Iris Murdoch. *Iris Murdoch Newsletter* 13 (*IMN*13), KUAS30/13

———. 2019. *Iris Murdoch*, Writers and Their Work. Liverpool: Liverpool University Press.

———. 2021. A Poor Player That Struts and Frets Her Hour Upon The Stage?: Iris Murdoch and the Theatre. *Iris Murdoch Review* 12 (*IMR*12): 36–48

Siddall, Stephen. 2009. *Landscape and Literature*. Cambridge: Cambridge University Press.

Silesius, Angelus. 2011. *Die Rose ist ohne warum*. In *Angelus Silesius*, ed. Gerhard Wehr. Wiesbaden: Marixverlag.

Stephen, Leslie. 1909. *Hours in a Library*, new ed., 3 Vols. London: Smith, Elder & Co.

Walmsley, Robert. 2016. *A Short History of the Cherwell*, 29 September 2016. https://cherwell.org/2016/09/29/cherwell-history-preface/. Accessed 4 January 2021.

Walton, Samantha. 2021. Four Apocalypse Sonnets. *bath magg* 5, https://www.bathmagg.com/samanthawalton/. Accessed 20 January 2021. Reproduced by Kind Permission of the Author.

White, Gilbert. 1853. *The Natural History of Selborne* (1789), ed. William Jardine. London: Nathaniel Cooke.

White, Frances. 2014. Raids on the Inarticulate: Poems for Wallace Robson. *Iris Murdoch Review* 5 (*IMR*5): 8–16

Wordsworth, William. 2007a. 'The Wanderer', Book I of *The Excursion* (1814), ed. Bushell, Butler, Jaye et al. London: Cornell University Press.

———. 2007b. *Lyrical Ballads*. ed. Michael Mason. New York: Pearson Longman.

Yeats, W.B. 2001. *The Major Works*. Ed. by Edward Larrissy. Oxford: Oxford University Press.

3

The Emerging Animism in the Novels: *The Flight from the Enchanter, An Unofficial Rose, The Unicorn* and *Nuns and Soldiers*

'THERE'S SOMETHING SPECIAL ABOUT *EVERY* STONE'

At the centre of Iris Murdoch's ecological consciousness is her apprehension of the equity of all life. Lucy Bolton says that Murdoch's 'equitable consideration of lives, female, male, animal, and vegetal', is 'one of the most refreshing and enjoyable aspects of reading Murdoch's philosophy' and, in the twenty-first century, she believes that it is right to view Murdoch as an intersectional feminist (2022, 444).[1] Bolton directs her readers to a key passage in *Metaphysics as a Guide to Morals*:

> Our pilgrimage (in the direction of reality, good) is not experienced only in high, broad, or general ways (such as in increased understanding of mathematics or justice), it is experienced in all our most minute relations with our surrounding world, wherein our apprehensions (perceptions) of the minutest things (stones, spoons, leaves, scraps of rubbish, tiny gestures, etc.) are also capable of being deeper, more benevolent, more just. (*MGM* 474)

---

[1] Intersectional feminism acknowledges multiple sites of identity, is inclusive in its demand for equity and endeavours to account for the different ways oppression and discrimination are experienced; see Crenshaw (1995).

© The Author(s), under exclusive license to Springer Nature Switzerland AG 2025
L. Oulton, *Iris Murdoch's Wild Imagination*, Iris Murdoch Today,
https://doi.org/10.1007/978-3-031-87833-6_3

Engaging the lexicon of pilgrimage, Murdoch conveys what she sees as a crucial but unrelenting moral quest to attend to and love the living world. Her just and benevolent view counters the devastating power dynamics of those who continue to advocate for human primacy over a suffering Earth. In this sense, Murdoch's approach to the natural world, like her feminism, comes from an intersectional position when she draws her reader towards a conception of environmental justice that takes account of not just human but more-than-human life.[2] As her fiction evolves, her portrayal of the natural world responds to something approximating Martin Heidegger's 'worlding world' where he suggests, 'If we let the thing be present in its thinging from out of the worlding world, then we are thinking of the thing as thing' (1971, 181). Heidegger's invented 'worlding' speaks to an ever-unfolding generative process that animates the world around us. Over the course of her novel-writing career, Murdoch's sense of equity slowly begins to emerge through her growing understanding of a participative, animistic world that, for those who pay attention, engenders equity and respect.

This chapter begins by examining some of Murdoch's earliest depictions of the natural world in her novels in order to trace the genesis of an emerging animistic sensibility that becomes increasingly striking in some of the later and more expansive fiction. Murdoch's close attention to nature's smallest details is evident from the start of her writing life. The profound sense of place and belonging for one character in *The Flight from the Enchanter* (1956) can be attributed to such experience. With this novel and with *An Unofficial Rose* (1962) we begin to see what is to become a pattern of solitary moments of epiphany in nature, a trait of a number of the later novels, and certainly of all the novels featured in this book. These become occasions of unselfing in nature when a form of momentary clarity is bestowed upon the solitary person, which Murdoch connects to love:

---

[2] 'Intersectional environmentalism', explains climate activist Leah Thomas, 'is an inclusive version of environmentalism that advocates for both the protection of people and the planet. It identifies the ways in which injustices happening to marginalized communities and the earth are interconnected. It brings injustices done to the most vulnerable communities, and the earth, to the forefront and does not minimize or silence social inequity'; see *London Environmental Network* (2022).

> The direction of attention is, contrary to nature, outward, away from self which reduces all to a false unity, towards the great surprising variety of the world, and the ability so to direct attention is love. […] It is in the capacity to love, that is to *see*, that the liberation of the soul from fantasy consists. (OGG 354)

Of course, Murdoch's reference to 'nature' here is to be understood as human nature. These moments of clear vision are often portrayed as short-lived and, in some cases, are shown not to have lasting effects.

In this chapter, I consider the garden as aesthetic space in its interchange with nature. I discuss Murdoch's more conventional use of the vegetal as symbol or motif in *An Unofficial Rose* (1962) before going on to investigate the marked divergence in setting and treatment with her conception of an agentive animistic landscape that emerges in the novel that follows, *The Unicorn* (1963). Later, as the novels become more expansive, so begins Murdoch's accommodation of vaster scales of space and time that counterbalance the minute details of the natural world that she observes, and works with, in the universe of her imagination. In some of these later novels, her representations of agentive primordial landscape, of vegetal and lithic materiality, become increasingly important and I consider all of these themes in the context of *Nuns and Soldiers* (1980). Murdoch's emerging animistic sensibility confirms the novelist's conception of an equity of all life that, in turn, has the potential to contribute to contemporary ecological discussions. This potential represents a hitherto unexplored element of a broader Murdochian ethics. Here, I connect Murdoch's conception of animism to expressions of ethical relation to the more-than-human through beauty, epiphany and justice—all relating to her evident love of, and respect for, the natural world.

Murdoch recorded observations she made in her own garden at Cedar Lodge in Steeple Aston, in nearby woodland, of the biota of the River Thames, of landscapes elsewhere in Britain, Ireland, and other countries she visited; these become a more significant feature of her personal journals over time. The entries where she remarks on the natural world, such as the one below, seem personal and contain fine sensory detail. On occasion, she would add observations to letters, too, and these, as might seem obvious, are written rather more with their intended reader in mind. For example, in a letter in early April 1965 to her friend and one-time lover Brigid Brophy—the brilliant polemicist and campaigner, with whom she particularly enjoyed being subversive—Murdoch attributes her feelings of depression not only to being unable to see Brophy, but to the 'absolute plague of daffodils in the garden' (*LoP* 290).³ Murdoch grumbles, 'I simply cannot share Wordsworth's feelings. (Well, I suppose his were wild ones, which might make a difference)' (*LoP* 290). To Raymond Queneau on 13 June 1968, she writes: 'The sun shines on Steeple Aston (it really does for once), I try to work in an intoxicating atmosphere of honeysuckle. I grow older' (*LoP* 367). To Philippa Foot, by this time living in Los Angeles, on 17 October 1982 she describes the Scottish landscape where she has just been on holiday with the Bayley family: 'my favourite sort of country (except not quite by the sea) with rivers bounding over boulders and descending waterfalls' (*LoP* 496). By contrast, her journal notes from 1989 comprise a more personal, reflective tone and seem to express more intimately an immersive appreciation of the natural world:

16 May
   Cow parsley viewing along the Cherwell.

18 May
   Sun, no wind, quiet air. Walk in the 'great crowd' at Wytham [Wood].⁴ Huge wonderful presences of trees. Vista of tree trunks, green, light beyond green. Empty, no one. Intensely beautiful and frightening and magical.

---

³ Horner and Rowe record that 'Brophy and Murdoch became very close during the 1960s and their love affair had a strong intellectual and emotional impact on Murdoch's life'; they suggest, too, that 'Brophy was a kind of alter ego for Murdoch, who relished and encouraged her friend's unconventionality' (*LoP* 603, 238).

⁴ It is poignant to read that Bayley continued to take Murdoch to walk in Wytham Wood in 1997. Stricken with Alzheimer's disease, Murdoch was reduced to watching *Teletubbies* on the television, a particular episode reminding Bayley of viewing the bluebells at Wytham Wood: 'They live in a thick and distant part of the wood, under dark conifers which stretch away downhill, and as they recede into darkness they light up into their most intense colour. [...] On the way there are real trees. Two gigantic sycamores, overpowering as a cathedral. But Iris has now a great fear of trees and I hurry her past them', John Bayley (1999, 244); Wytham Wood's ancient woodland is now a Site of Special Scientific Interest, belonging to the University of Oxford, and can be visited by arrangement.

> 23 May
>   Wonderful swim in Thames. Oxey Mead, Witney Road. Earliest? Hawthorn, buttercups, orchids, comfrey, larks, reed warblers, cuckoo. […] Glossy water. (J14 111)

These acute observations she notes down are fundamental to the work of Murdoch the poet and fiction-writer.

The intensity with which she attends to the 'minutest things' (*MGM* 474) is evident from the inception of her writing life, as conveyed by the poem of 1952 discussed in the previous chapter, 'The trailing stars tell of dooms' (PN6 12 and *IMR*5 13). The speaker of the poem glimpses 'the fall of the world / Poised at the intricate centre of flowers', and the way in which a lamp lights up in the dark depths of a suspenseful cat's 'enormous eye' (PN6 12 and *IMR*5 13). The words draw readers in to tight focus, as if invited to peer through a microscope on an intensely detailed observation. In a scene belonging to Murdoch's second novel published just four years after writing this poem, we witness again the same sort of intense attention to nature's minutiae.

The Flight from the Enchanter ostensibly addresses the social dislocation of war and the consequent loss of rootedness and belonging for its refugees and other victims. By stark contrast, British civil servant John Rainborough has been able to live his entire life in the large family home: 'a safe stronghold' in London 'just off Eaton Square', where 'a dreamy wistaria has been growing for several scores of years' (*FFE* 118, 121). The wistaria symbolizes Rainborough's own firmly rooted life-experience, yet Murdoch also succeeds in capturing a sense of the climber's reality and its intrinsic value in creating Rainborough's sense of belonging:

> This wistaria was connected in Rainborough's mind not only with his childhood but with what he regarded as all his deepest thoughts: those phantoms through whose nebulous forms, as through the bodies of ghosts, he had seen, sitting for hours on end at the drawing room window, the knotted branches and the feathery leaves, until the outer world had disappeared altogether, mingled with thought and transformed into an inner substance. (*FFE* 121)

Murdoch conveys the sensory experience of the warm spring day that comforts Rainborough with an 'image of summer' (*FFE* 123), conjuring the deeply held memory of all his summers. Akin to the 'ghostly walkings' (*NG* 332) that Mary Clothier feels through her body in *The Nice and the Good* which I discuss later on, here Murdoch creates a multi-layered kaleidoscope of images, real, imagined, ethereal, intangible, to suggest how outer experience of landscape can fix memories in the inner life of the mind. In doing so, she expresses the significance of the sort of embodied experience of place of which refugees and other displaced people are all too often cruelly deprived. Rainborough, however, is seeking short-lived (as it turns out) solace in his garden, personally threatened, he feels, by the onward march of progress which sees not only increasing numbers of women like Agnes Casement in the workplace but also the necessary requisition of a part of his family garden for the neighbouring hospital's new x-ray department.

The garden has meant security and constancy for the solitary Rainborough and it is in this moment that Murdoch conjures her image of 'minute relations with the surrounding world' (*MGM* 474), involving, again, an intense scrutiny of the 'intricate centre' (PN6 12 and *IMR*5 13) of flowers:

> Hyacinths, narcissi, primulas, and daffodils stood before him, rigid with life and crested with stamens, tight in circles, or expanding into stars. He looked down into their black and gold hearts; and as he looked the flowerbed seemed to become very large and close and detailed. He began to see the little hairs upon the stems of the flowers and the yellow grains of pollen, and where a small snail, still almost transparent with extreme youth, was slowly putting out its horns upon a leaf. [...] He looked at the snail. Can it see me? He wondered. Then he felt, how little I know, and how little it is possible to know; and with this thought he experienced a moment of joy. (*FFE* 123)

Rainborough's intense examination of the tiny creature will later be echoed in *The Sea, The Sea* when, in a 'glow of concentration', James Arrowby carefully captures a fly 'with a tumbler and a sheet of paper' to release it out of the window of his Pimlico flat (*TSTS* 173). Then, at

Shruff End shortly after, a fly perches on James's finger and Charles notes with fascination the interaction between the creature and his cousin: 'It stopped washing. James and the fly looked at each other' (*TSTS* 179). Murdoch's own concern for the smallest of creatures is evident in a letter sent to Brophy in autumn 1965 in which Murdoch despairs at the apparent hierarchy of concern for animals:

> How awful to think there are Venus flytrap fanciers at large amongst us. It's odd how people's sympathy with living creatures diminishes in relation to size (& also outside the mammalian world.) Even Wittgenstein knew that it's not difficult to attribute pain to a fly.[5]

In *The Flight from the Enchanter*, Rainborough is momentarily captivated by the tiny and intricate details he sees and more generally humbled by the family garden.

The scene represents an early depiction of a solitary character experiencing the momentary ecstasy of pure joy in nature, an occasion Murdoch will reprise in many of her later novels. Patricia Duncker recognizes this moment as 'the classic experience of mystical contemplation' (2000, xi). Murdoch captures 'the moment', she says, 'wherein the separation of outer and inner substance, mind and world, ceases to exist and we become one with all that is' (Duncker 2000, xi). The profound effect on

---

[5] Letter to Brophy from Murdoch, KUAS142/5/143, in the Iris Murdoch Collections; in her letter, Murdoch thanks Brophy 'for animals'; this is Brophy's 'The Rights of Animals', first published in *The Sunday Times* on 10 October 1965. Brophy must have sent her a copy. Brophy vehemently opposed bloodsports, vivisection and, indeed, anything that involved the exploitation of animals. In condemning vivisection, Brophy asserts: 'Biology [...] offers us no pretext for treating our own species as absolute and all the rest as slaves. Since the discovery of evolution, biology has been making it plain that the frontiers between species are flexible, that we indeed are animals. [...] Any improvement in the world's sensibility to rights is helpful to all beings who are oppressed. And conversely it is dangerous to argue that in order to find cures for human sufferings it is justifiable for vivisectors to cause suffering in other animals. A medical profession founded on callousness to the pain of other animals may eventually destroy its own sensibility to the pain of humans'. Brophy's essay 'In Pursuit of a Fantasy' was later collected into an anthology, *Animals, Men and Morals*, ed. by Stanley Godlovitch, Rosalind Godlovitch and John Harris (1971, 125–45). Murdoch alludes to Wittgenstein's 'Look at a stone and imagine it having sensations.—One says to oneself: How could one so much as get the idea of ascribing a sensation to a thing? One might as well ascribe it to a number!—And now look at a wriggling fly and at once these difficulties vanish and pain seems able to get a foothold here, where before everything was, so to speak, too smooth for it' (1958, Part 1, §284).

Rainborough—'he felt, how little I know, and how little it is possible to know' (*FFE* 123)—is to engender humility before these intricate details that brings him, if just for a moment, out of his brooding self-obsession. As Duncker has observed, the moment fittingly recalls Andrew Marvell's 'Annihilating all that's made / To a green thought in a green shade' (1996, 100). Considered the first poem in English about nature, Marvell's 'The Garden' pictures the 'Far other worlds' that are conjured through peaceful contemplation in such a place (Marvell 1996, 100–102, l. 47–48). Unfortunately, the novel's short-lived and touching garden scene provides a conspicuous counterpoint to the overarching menace of a novel set to depict the 'casual slaughter of fragile, living things' (Duncker 2000, xi)—Mischa Fox has killed a kitten (*FFE* 208), Rainborough crushes a handsome moth (*FFE* 137) and even fish suffer a slow death (*FFE* 196).

Rainborough's joyful moment of selfless contemplation is fleeting. Shortly afterwards he brutally assaults Annette and then, on being disturbed by Mischa's arrival, hastens to conceal the young woman in a cupboard. An odious exchange between the two men ensues during which Mischa declares how perfectly possible it is to create the ideal woman simply 'by breaking her—' (*FFE* 135). While Rainborough may appear appalled at the idea, the conversation only occurs in the first place because he has shared his frustrations with Mischa on the influx of women (a 'vast legion of clever and provoking females') to SELIB (the Special European Labour Immigration Board) who 'take one's work away' and 'make the place pointless by being there' (*FFE* 132). Rainborough is indeed complicit in the menace. Rainborough metes out violence on his once beloved garden too. On the eve of the planned demolition of both wall and wistaria, he takes matters into his own hands and, with Agnes Casement's help, destroys the climber and surrounding plants before construction workers can begin clearing the area for the new building. The garden is soon unrecognisable, reduced to 'piles of masonry […] brick-stacks, cement-mixers, and […] Nissen huts' (*FFE* 216), and no longer represents the stable force it once did in the volatile man's life.

Some critics appear to have expressed certain understanding for the civil servant's predicament. Conradi, for example, describes a 'sedate and unremarkable' Rainborough involved in 'various arbitrary violences' but, notably, does not mention the civil servant's sexual assault of Annette

(*S&A* 68, 67). Where a contemporary reader may more readily align with Conradi is in his suggestion that *The Flight from the Enchanter* meditates on 'pity and power' (*S&A* 68). Nonetheless, Rainborough's garden—part realism, part metaphor—clearly represents an important site of contemplation and memory and, at the same time, symbolizes the unravelling of everything the civil servant, in an age of momentous change, has held to be true. The garden pictures place as intrinsically valuable to a person's sense of identity, dignity and well-being, and juxtaposes the momentous challenges faced by those characters who have been uprooted and feel dispossessed, and the general menace of casual cruelty in the novel.

'Love Accepts the Contingent'

At this early stage in Murdoch's fiction-writing life she is yet to truly imbue the garden, indeed any manifestation of the natural world, with a life force all of its own. The significance of a garden's hybridized space, for Mateusz Salwa, is that it is 'where nature becomes culture, and culture becomes nature' (2014, 317). Gardens are places, he argues, 'where we somehow experience nature aesthetically' and, 'it is precisely the aesthetic attitude which is inherent to gardens that makes one approach nature present in them in ethical terms, as well' (Salwa 2014, 317). Hybridity of this sort has a significant role to play in Murdoch's sixth novel, *An Unofficial Rose*.

In *An Unofficial Rose*, the realist landscape bears certain precision: Grayhallock is located on a hillside above the 'flat pale green expanse' of Romney Marsh in Kent, twenty miles from the 'invisible' sea in England's South Eastern corner (*UR* 18). The novel, notable for its tight framework and stylized use of symbol, was recognized at the time of its publication for taking formal inspiration from Henry James. In his contemporaneous review, however, Roger Sale suggests that the Jamesian influence only goes so far: 'Everyone in an Iris Murdoch novel has a room that is distinctively his own, but the loving descriptions of these have little of […] James' symbolic evocations' (1963, 424). Byatt agrees with Angus Wilson on Murdoch's tendency to verge 'on an unconscious parody' (1994, 139) of James in *An Unofficial Rose*. Murdoch, herself, admits to 'a good deal of affinity' with James in an interview with W.K. Rose first published in *Shenandoah* in 1968 and it is here that she publicly acknowledges his formative influence on her writing (*TCHF* 28). In a recent podcast to mark the novel's sixty years, Frances White notes strong formal, thematic and intertextual links to James, and to Jane Austen, that are wholly intentional in her view. Shrouded in a 'distanced narrative perspective', White argues that, at the time of publication, critics did not realize Murdoch's 'deliberate placing of artifice', in effect, superimposing the mannered tradition of the 'English tea-party novel' onto a fictional work of the early 1960s (*Podcast* 6 December 2022). Murdoch makes full use of imagery in this novel and Rowe is another critic who connects Murdoch to James in this regard. She notes how Murdoch (in Jamesian terms) wanted to 'explore how the "sister arts" could "sustain and explain" each other' (Rowe 2019, 78, cites James 1987, 188). *An Unofficial Rose* is one of her novels that is 'saturated in painterly imagery', Rowe remarks (2019, 81). Nonetheless, Murdoch was wary of symbol. She felt that the 'aesthetic centre it represent[ed]' should be 'resisted by a competing force within characters', according to Conradi (*S&A* 120). He appears to suggest that one should be aware of some sort of pushback or ambiguity in interpretation. Thus, it seems that Murdoch had her own *modus operandi* in determining their use.

There are two central symbols in the novel—the first, the widely discussed fictional Tintoretto study, and a second, the dogrose *Rosa canina*. The attention given by early critics of the novel to the first has served to overshadow the significance of the second and, in doing so, has eclipsed the dogrose's potential in a broader ethical interpretation of the work. The Kentish landscape of *An Unofficial Rose* does not project the sort of richly pulsating animism of later novels such as *Nuns and Soldiers* or *The Good Apprentice*, nor as readers are we given a sense of an embodied world either: the world perceived through the body of individual characters that can be richly experienced in reading Murdoch's later novels. What we do encounter again, in reading this novel, is the affective response of a solitary character in landscape. Unlike later novels, where Murdoch succeeds in animating the multiplicity of life beyond her human characters, *An Unofficial Rose* seems to be far more tightly constructed around the symbolic use of its visual elements.

Murdoch conjures elements of the natural world in her fiction in much the same way in which she engages with visual art. In *The Visual Arts and the Novels of Iris Murdoch*, Rowe classifies both the novelist's 'detailed references to the natural world' and the works of art that tend to feature in the novels as 'visual phenomena' (2002, 62). Rowe discusses how these images might create meaning and provoke a response: 'Images work in two ways: the reader observes how they influence characters while they are themselves under a similar influence', that is, 'readers unconsciously assimilate images which influence value judgements' (Rowe 2002, 62). Such images serve to communicate with the reader independently of language, and we are thereby reminded of Murdoch's assertion in 'Thinking and Language' about the indefinability or inseparability of language, imagery and sensation:

> We naturally use metaphors to describe states of mind, or to describe 'thought processes', in those cases where a sentence giving the verbal content of the thought is felt to be inadequate. In such a context metaphor is not an inexact *faute de mieux* [for want of a better alternative] mode of expression, it is the best possible. Here metaphor is not a peripheral excrescence upon the linguistic structure, it is its living centre. (TL 39–40)

Thus, for Murdoch, metaphor becomes the 'living centre' (TL 40) of communication when words fail; thinking is not just a means of 'designating', but it is about 'understanding, grasping, "possessing"' (TL 41). For *An Unofficial Rose*, the success or otherwise of the storytelling relies on a fictional work of art on the one hand and, significantly, a species of wildflower on the other.

The privately owned Tintoretto study, hanging in Hugh Peronett's London flat, is regarded by some critics as the key image: the story 'might be said to centre around the Tintoretto painting owned by Hugh Peronett', says Byatt (1994, 137). Rowe is more precise about its significance, suggesting the 'novel constructs a dialogue with Tintoretto's *Susannah Bathing* to illustrate a type of male consciousness that purposefully transforms female beauty into eroticized desire' (2019, 85). The 'glowing Tintoretto' inherited by Hugh from his late wife's family represents a source of erotic fantasy for him, and a 'shrine of refuge' away from 'Ann's worthiness, Randall's sulks, Swann's piety, Miranda's pranks, Penn's accent, and the gawky rows of dripping rose bushes', Murdoch tells her reader (*UR* 49). However, and uniquely it seems in a Murdoch novel, the artwork (as viewed by Hugh's son Randall) also represents a financial asset, and its sale becomes Randall's obsession when he needs money to leave his wife, Ann, relinquish the rose nursery business and flee with his lover, Lindsay Rimmer, to a life of luxury.

Rowe asserts that the Tintoretto study, an entirely invented precursor to the world-renowned *Susannah Bathing*, is an important reason why the novel as a whole is 'less than satisfactory' (2002, 75). She argues that its representation is the 'most sustained attempt at using a painting as a central, extended metaphor for an erotic servitude that remains forever outside the reaches of the Good', which suggests, in her view, that a detailed knowledge of Tintoretto in general, and *Susannah Bathing* in particular, is necessary to make the novel work successfully (Rowe 2002, 62). The Tintoretto does not function as a source of revelation such as that experienced by Dora in *The Bell*, in the public realm of the National Gallery before the portrait of Gainsborough's daughters. On the contrary, the study becomes 'transformed by the ego into its own private fantasy' (Rowe 2002, 78) by Hugh and Randall, in the private realm of Hugh's London flat. The artwork is fetishized: who owns it and, indeed, who is looking at it are significant. Viewed as the novel's central image, the fictional Tintoretto appears to have encouraged critics of *An Unofficial Rose* to focus on the choices confronting two generations of men that, in turn, have engendered objectifying discussions that point to dull wives on the one hand, and more exciting mistresses on the other.

Significantly, however, Murdoch's novel sustains another important image, the dogrose, in the wild and unpossessed. If Hugh's Tintoretto functions 'outside the reaches of the Good' (Rowe 2002, 62), then the wild dogrose of the novel's title is emblematic of the wild rose from which all roses have at one time been grafted, hybridized and cultivated and it attempts to picture the Good. Yet, as the pivotal image of the novel, the 'unofficial rose' seems to have gained relatively little attention from critics. Dipple is one who does acknowledge its significance in the centrality of Ann:

> The structure pivots around Ann's character, and that the simple English dogrose, the unofficial rose, rather than the impressive hybrids the Peronett

nursery (the world) produces, should signalize this odd-girl-out, focuses the conclusion sharply. (1982, 59)

However, sensing a weakness in Ann's character—'Ann is impotent throughout'—Dipple does not think the novel successful for it (1982, 59). However, it is my contention that in foregrounding an aesthetic attitude to natural beauty, Murdoch is placing Ann at the novel's moral centre and, when considered in this way, a different ethical perspective to the novel and its characters comes into view.

Unlike the demands made of her reader by the great visual art that Murdoch engages with in her fiction, botanical allusions require arguably less intellectualizing from a reader attempting to grasp what is at stake. The natural world draws a character's (and therefore a reader's) attention to a clearer view of the world outside themselves: botanical elements present an opportunity for moral progress away from self-obsession and delusion and a tentative step towards the Good. The potential of such things, in capturing our attention, can engender a feeling analogous to love.

Murdoch has prioritized the botanical image by her choice of title for this novel. By adopting the line from Rupert Brooke's 'The Old Vicarage, Grantchester', the scrambling dogrose, the 'unkempt' beauty of the hedgerow, becomes her subject and will contrast in stark terms with the manufactured rose cultivars that feature in the story:

> Unkempt about those hedges blows
> An English unofficial rose;
> And there the unregulated sun
> Slopes down to rest when day is done. (Brooke 1992, 260)

Brooke's ironically conceived poem of place—at once both comic and cruel—rehearses the poet's ambivalent nostalgia for the English countryside. Brooke composed his poem in the *Café des Westens* in Berlin, a city where, he claims, 'tulips bloom as they are told' (1992, 259). The poem expresses a deep longing for the natural beauty of the English countryside which is, for Brooke, contrastively synonymous with freedom. In adopting Brooke's line for her title, Murdoch succeeds in foregrounding the pale and understated natural beauty of the dogrose scrambling in the hedgerows, all the while Grayhallock's specialist rose nursery grows its hybrids and cultivars in regulated rows.

The rose nursery, established by Randall Peronett, is run (it seems along with everything else) by his long-suffering wife Ann, and it is Randall, himself, who aligns Ann with the dogrose. Although he has become disenchanted with his business, he remains proud of what he sees as 'his own creations' (*UR* 31), naming cultivars after family members and prizing the perfection of form he claims proudly to have manipulated into his masterpieces. It is not difficult for some readers to be seduced by Randall's artistry and ego.

Dysfunction in the marriage, then, is conveyed taxonomically. Randall is dismissive of Ann to his father, Hugh: 'It's what she *is* that does it […] She's as messy and flabby and open as a bloody dogrose' (*UR* 32). For Randall, the dogrose represents the untameable mess of the contingent world. 'I need a formal world. I need form', he declares, clutching at the rose that his father has absently removed from a nearby rose bowl: 'Yes, yes, form, structure, will, something to encounter, something to make me *be*. Form, as this rose has it. That's what Ann hasn't got' (*UR* 32). Ann and Randall's grief for their dead son, Steve, has put tremendous strain on the marriage:

> Randall had been reeling drunk at [Steve's] funeral and said later that Ann 'had never forgiven him'. It was rather perhaps that he had never forgiven Ann, upon whom by some insane and fantastic logic he had seemed to fix the blame of his bereavement. (*UR* 13)

From Hugh's interiority, revealed to readers during his wife Fanny's funeral, we learn that Steve possessed a charm like his father, a charm that makes them lovable in the way that 'the worthy and deserving ones, such as Ann, are, by a terrible justice, unloved' (*UR* 14). Yet, the bereaved Ann had been nursing mother-in-law Fanny, looking after Grayhallock and running the rose nursery while coping with the unfaithful Randall. By outlining the couple's differences and feeding Randall's frustrations to her reader, Murdoch might appear to be luring her reader into taking sides, even into condemning Ann.

In a television interview, Murdoch tells Frank Kermode that she did not think of *An Unofficial Rose* as a particularly successful novel (1965). As one of Murdoch's good characters, Ann has certainly succeeded in frustrating critics: early criticism of her is excoriating. Dipple says Ann is 'vague and structureless' from the way she mourns her mother-in-law, Fanny Peronett, to 'the rest of her activity' which renders her 'impotent' (1982, 59). Dipple is unconvinced by Ann's good character. She says: 'her failures of energy combine with wistfully bitter resentments because she lacks the machinery of will' (1982, 60). She concludes that Ann is 'indeed deadening, and Randall is in this respect right about her' (1982, 58). That Murdoch would wish her reader to sit in judgement in this way remains far from compelling.

Criticism in the twenty-first century offers new perspectives on the novel and on the character of Ann. Rowe makes the important observation: 'Critics largely failed to notice that Murdoch was experimenting radically with finding new ways of illustrating the interiorization of unhealthy masculine perceptions of women that cause self-serving and callous behaviour towards them' (2019, 17). In the podcast, I point to a passage that appears to suggest that Ann is, in fact, a victim of emotional abuse (*Podcast* 6 December 2022):

> The particular quality of her long battle with Randall had seemed progressively to empty the certainties by which she lived, as if the real world were being quietly taken away, grain by grain, and stored in some place of which she had no knowledge. This did not make her doubt the certainties. There would be for her no sudden switch of the light which would show a different scene. But there was a dreariness, a hollowness. She could not inhabit what she ought to be. (*UR* 107)

Bit by bit her sense of self has ebbed away, defeated by the behaviour of her fantasy-driven husband. Ann has been awaiting a visit from the pious Reverend Swann who greets her with 'an intense look of wordless sympathy' (*UR* 107). He is there to remind Ann of the sacrament of marriage, intoning: 'We must not expect our lives to have a visible shape. They are invisibly shaped by God. Goodness accepts the contingent. Love accepts the contingent. Nothing is more fatal to love than to want everything to have form' (*UR* 110). Later, he attempts to impose far too much pattern and influence on Ann: 'We know, don't we, about the permanence of marriage?' (*UR* 229). Ann knows she has little choice but to 'set Randall free' (*UR* 229). Larger-than-life and faintly comic characters—such as the Reverend Swann and Emma Sands, writer of murder mysteries, former lover of father-in-law Hugh and erstwhile companion of Randall's new lover ('All I think of when I see something beautiful is how to make it an occasion of violent death' [*UR* 132])—help to render Ann's own character apparently weak, unobtrusive and possibly dull.

In 'The Idea of Perfection', in her famous 'M and D' exemplar, Murdoch encourages her reader not to judge harshly but to engage in an 'endless task', to look again (IP 317). A mother (M) feels a certain animosity towards her daughter-in-law (D) because D appears 'vulgar', 'noisy', 'undignified' and 'tiresomely juvenile' (IP 313). Murdoch says that 'M tells herself: "I am old-fashioned and conventional. I may be snobbish. I am certainly jealous. Let me look again"' (IP 313). After deep reflection, however, M begins to see D as 'refreshingly simple', 'spontaneous', 'gay' and 'delightfully youthful' (IP 313). '[H]er outward behaviour, beautiful from the start, in no way alters' (IP 313). The transformation has happened entirely in M's mind, where she has been 'morally active' (IP 314). Which leads one to question what might happen if we were to look again upon Ann 'justly and lovingly'? (IP 317). It also raises the issue of how a novelist might convey the ruminations belonging to the inner life of a character. By connecting Ann to the understated pale-pink hedgerow rose and using Randall to form this connection, Murdoch presents an altogether different ethical picture. In 'The Fire and the Sun', Murdoch writes:

> Virtue in general may not attract us, but beauty presents spiritual values in a more accessible and attractive form. The beautiful in nature (and we would wish to add in art) demands and rewards attention to something grasped as entirely external and indifferent to the greedy ego. (FS 417)

Ann's understated virtue, of itself, may not immediately attract us. What ought to draw readers to look again is Randall's alignment of Ann with the *Rosa canina*, the 'beautiful in nature' (FS 417).

When Randall succeeds in banking the proceeds from the sale of the Tintoretto, he is overcome by a strange set of emotions. He almost feels as if he has killed his father, has no desire to see his lover Lindsay, but has the 'luxurious [...] sense of having [her] *stored away*' (*UR* 168). He feels an 'agonizing sense of exclusion' that life at Grayhallock continues without him (*UR* 182). He returns for one 'last look at the roses' and to face what Murdoch describes as his 'moment of annihilation' (*UR* 182–83). He fancies that, as he turns his back on Peronett Rose Nurseries, it 'would cease to be as completely as if they had been sunk in the Marsh' (*UR* 183). His self-delusion causes him to believe that his floral creations would live on as 'purer distillations of his being, when their namesakes were only so much manure' (*UR* 183). But he feels done with what he sees as the 'vulgar pursuit' of the 'endless tormenting of nature to produce forms and colours far inferior to the old and having to recommend them only the brief charm of novelty' (*UR* 183). He cannot conceive of the nursery surviving without him.

Randall's self-pity is suddenly interrupted by a realization: 'The true rose, the miracle of nature, owed nothing to the hand of man' (*UR* 183). Randall loves the old roses, becomes 'completely absorbed into a heaven of vision', and is overwhelmed by their 'transcendent purity' (*UR* 183). Murdoch's narrator says,

> The old roses [...] made him, for the moment, like to themselves. He could have knelt before these flowers, wept before them, knowing them to be not only the most beautiful things in existence but the most beautiful things conceivable. God in his dreams did not see anything lovelier. Indeed the roses were God, and Randall worshipped. (*UR* 183–84)

The roses gain agency as the solitary Randall submits, his ego momentarily annihilated. He is enveloped in their scent and looks 'with ever new amazement' at their exquisite form and 'those formulae that Nature never forgot' (*UR* 184). In this moment of high eros, Randall encounters Nancy Bowshott, his loyal employee, who pleads with him to take her with him: 'He saw her, but only for a second, as a separate being with troubles and desires of her own [...]. She seemed to him with her flushed face and her fierce brow and her disordered hair suddenly beautiful' (*UR* 185). For a brief moment he sees her as quite real and separate from himself. Then the old Randall kisses her savagely and the roses, we are told, '[fall] to the ground' (*UR* 185). Transitory and fleeting, the revelatory moment has vanished.

Later, in Rome with Lindsay, Randall awakes to 'the sounds of the early morning household', believing himself to be at Grayhallock (*UR* 258): 'He had never dreamed so much in his life', and '[e]very night in dreams he saw Ann', and never Lindsay (*UR* 259). He acknowledges:

> There was only one thing in the world that he was really good at, and that he would never do again. He saw in a vision the sunny hillside at Grayhallock with its slight haze of green and its myriad little coloured forms and he sighed. Ann. (*UR* 263)

Just as the *Rosa canina*, the dogrose, is crucial for the grafting or budding of cultivated roses, we are left to wonder whether Randall may prove ultimately unable to live without Ann.

The novel closes, as it opened, with a means of entry into Hugh's interiority. He is gratified that 'like a wild bird' his son is free and that Ann 'seemed hardly to have noticed Randall's departure so little difference had it made to the shape of her days' (*UR* 285). It seems that Lindsay's function was to free Randall from Ann. And yet, Randall 'could always, and after his own beautiful fashion, return to Ann. Ann would always be waiting' (*UR* 263). The roses provide momentary epiphany for Randall; the flowers themselves briefly gain agency, yet there is little to prefigure the animistic qualities of later novels. Remarkably, just one year after the release of *An Unofficial Rose*, Murdoch publishes a novel wherein her treatment and the participation of landscape has undergone striking transformation.

'Love Holds the World Together'

If gardens provide a means of looking at nature in an aesthetic mode then Murdoch's depiction of bleak wilderness contrastively expresses the natural world's contingency, its numinous agency, its sublimity and autonomous presence. Murdoch's seventh novel, *The Unicorn*, reveals a step-change in the novelist's treatment of landscape and heralds an emerging engagement with animism. Animism is an ancient means of trying to make sense of a dynamic and participatory world. Edward Tylor, the Victorian ethnographer often regarded as the founder of anthropology, reportedly thought of animism as a sort of mythopoetic belief system handed down through generations and as a crudely reverential precursor to established religion.

According to Graham Harvey, the idea of a sort of naïve animistic belief system evolving into more sophisticated systems of faith represents what is now known as the 'old animism'; it is seen as the 'allegedly unwarranted attribution of life to objects and personhood to nonhumans' (2017, 206). Harvey explains that, conceptually, 'new animism' is 'less about attributing life and/or human likeness, than it is about seeking better forms of personhood in relationships' (2017, 16). In the preface to the second edition of *Animism: Respecting the Living World*, Harvey says that 'person', 'relative' and 'relation', for him, are synonymous terms when discussing the more-than-human world, yet his use of 'person' and personhood' seems to have been provocative for some (2017, xiv). He concedes that on reflection he could perhaps have used 'person' less and relative/relation more in his animism discourse (Harvey 2017, xiv). Yet, ascribing personhood to non-human objects is nothing new: corporate entities, in the United Kingdom at least, have had legal personhood and been able, for example, to own property since the 1800s. Indigenous groups have led the way in assigning personhood status to natural entities and this has become, for instance, about how we protect rivers and forests and other entities in the natural world, in their own right as well as for the collective benefit.[6] New animist theory is less about attribution and more about valuing reciprocity and respect. It challenges the Cartesian separation of consciousness and matter, mind and body, nature and culture/society. It coalesces with modern conceptions of ecology and represents a way of life for some including a host of pagan, indigenous and other peoples who might engage in activism but more often express their beliefs 'in relatively simple, everyday respectful behaviours that treat the

---

[6] For a recent assessment of legal personhood, see Becky Ripley and Emily Knight (2023) 'Rivers and the Rights of Nature'.

world as a diverse and vibrant community of persons' (Harvey 2017, 210). In this sense, new animism aligns with ideas about the consciousness of matter and how consciousness resides in the sensing human body.

The young Murdoch possessed a sense of animism that felt deeply personal and proved life-long. On 4 June 1945, she observes in herself an 'animism—embracing cats, buses, stones… A tenderness for all things. Such as my mother has' (J3 13). In another journal entry, on 21 February 1949, Murdoch records that she had a

> curious hesitation today about burning a sheet of paper. There is a sort of animism which I recognize in myself & in my parents. We are surrounded by live and rather pathetic objects. I connect this with my sentimentality & general softness. (J7 5)

Conradi connects her animism to 'an extreme protective anxiety about her world and all it contained' (*IMAL* 276). Frances White remarks on a 'poignantly personal sense of animism' that 'permeates Murdoch's work, both philosophical and literary' (2020, §2). In terms of her fiction, Murdoch begins to put her animist affinities on show in her seventh novel, *The Unicorn*, where she displays a certain commitment to characters for whom 'consciousness is inescapably embodied, mattered and performed' (Harvey 2017, 193) within the context of their equally significant surroundings. Animist landscapes achieve greater significance in later novels, where episodes are no longer seen to be exclusively triggered or bounded by characters but may equally be instigated by the enchantment of structure or of landscape.

Enchantment is afforded significant exposition in Murdochian fiction. Conventionally, readers of Murdoch's novels tend to think of enchantment as emanating from particular charismatic characters. For instance, Mischa Fox in *The Flight from the Enchanter* and Honor Klein in *A Severed Head* (1961) are considered enchanters. In the case of *The Unicorn*, the mysterious inhabitant of the suitably named Gaze Castle, Hannah, is the character who projects the 'enchanter's most enigmatic and compelling aspect' which Linda Kuehl defines as 'a mythic dimension symbolically conveyed through [her] gaze' (1969, 350). However, as Patrick Curry avers, enchantment need not only be attributed to the person;

3 The Emerging Animism in the Novels: *The Flight*...   115

enchantment is 'always both material *and* spiritual, precise *and* mysterious, limited *and* unfathomable' (2014, 469). Curry guides his reader to the etymology of the word 'enchant' which, via the French *enchanter*, is rooted in the Latin *incantare* (*in+cantare*), to sing (2014, 470). An enchanter can thus suggest something that actively transmits something of itself outwards, but not only that: enchantment that involves transmission has the potential to provide something immersive, potentially collaborative and mutually reactive.

With this understanding of enchantment, animism becomes the name for the meeting point of different dynamic lives:

> Animist enchantment is strictly non-anthropocentric, so all kinds of beings, including 'things', can turn out to be existentially alive, and any object a subject with agency and an agenda, with whom one finds oneself in a relationship. (Curry 2014, 470)

In the same way that the natural world cannot be viewed merely as providing plenteous resources for human exploitation, neither can it be viewed as somehow separate and apart from us. Instead, as Donna Haraway advocates, we are living constantly surrounded by and immersed in 'a world full of cacophonous agencies' (1992, 297). *The Unicorn* is the first of Murdoch's novels to start to picture such non-human agencies.

Set in Ireland, *The Unicorn*'s 'appalling' landscape has been affected by a tumultuous weather event, in which 'a lake came flooding down the valley' (*TU* 7, 9). The perilous atmosphere of the opening chapters is initially created by the sense of an impending threat of a further weather event exacerbating the already devastated landscape. The wildness of the region seems particularly arresting after the sedate formality of the rose gardens in *An Unofficial Rose*. Conradi describes the wild terrain of Ireland's west coast in *The Unicorn* as 'a sea which kills people, rocks with carnivorous plants, [and] a featureless bog which picturesquely floods at seven-year intervals' (*S&A* 154). Yet, at the same time, he seems to limit the significance of the expansive and 'very ancient land' with megaliths (one of them 'seemingly pointless yet dreadfully significant' [*TU* 15]) to a sanitizing remark on 'the striking use of stage props and scenery of the Romantic sublime' (*S&A* 154). Such commentary suggests authorial contrivance and overlooks the animism and the agentive power of the virtually treeless peat bog and seascape engendered in the novel and fundamental to its unique character.

Murdoch engages with the remote Celtic setting and establishes a sense of unease and, at times, peril within the novel's narrative. She fictionalizes County Clare's Burren region, naming the vast bleak ancient limestone landscape the 'Scarren' (*TU* 12), and establishes the wild contingent space surrounding Gaze Castle, the main locus of the action. Avril Horner has examined the use of the Gothic in *The Unicorn* and suggests that Murdoch uses the landscape thematically to 'exploit and parody Gothic effects' (2010, 70); her purpose, she says, is 'to explore the possibility that evil is not just the absence of good, but a dynamic force to be recognized and rejected by those pursuing moral goodness' (2010, 71). Murdoch's intense interest in narrative and the problem of representation feed her compulsion to experiment with each new novel. Horner avers that Murdoch was also driven

> to keep questions of good and evil firmly within her creative and intellectual remit at a time when such terms were regarded in many quarters as absolutes (and therefore as invalid). Working with the Manichean world of the Gothic enabled her to playout in fiction her belief that there are values and realities that transcend humanity. (Horner 2010, 72)

The Unicorn's apparent instability and the suggestion that other forces are at play have helped generate discussion of Murdoch's purpose in her (self-conscious) invocation of the Gothic in this novel. Horner explains that the Gothic

> relies on sensationalism, melodrama, the irrational and excess for its effects. These effects are used to evoke emotions such as fear, terror, horror—and to provoke actual bodily responses (the hair rising on the back of the neck, the chill running down the spine). (2010, 71)

Everything in *The Unicorn* is 'menacing', a descriptor that 'tolls repeatedly throughout the novel', she says (Horner 2010, 77). It is one of a cluster of novels that feature Gothic elements, including precursors, *The Bell* and *A Severed Head*, and *The Time of the Angels* published three years after *The Unicorn*. Murdoch places all of them in the 'realm […] of the Freudian "uncanny", the phenomenon where something familiar becomes troublingly unfamiliar (or *vice versa*) when it occurs in an unexpected context', explains Bran Nicol (2004, 55). He notes that the 'geography' of *The Bell* 'plays on gothic conventions', and this is certainly true of *The Unicorn* too (Nicol 2004, 54). Underpinning *The Unicorn*'s Gothic mission is its setting: a large house surrounded by a reverberating force field of wildness.

Gothic's concern with the porosity of boundaries opens up the novel's interpretive potential when Murdoch successfully creates the 'condition of rupture, disjunction, fragmentation', considered characteristic of the genre (Miles 2002, 3). Key participants in the tale appear ill at ease. Marian is frightened by her surroundings and feels 'completely isolated and in danger' (*TU* 16), and Gaze Castle 'somehow, resembled her strangely, it was nervous too' (*TU* 29). Jamesie tells Marian about the man who sank and died in the peat bog: 'I wouldn't go up there myself in the dark for any money' (*TU* 46). Even the immensely self-confident but suggestible Effingham is perplexed by the exultant and mocking Pip Lejour's revelations of the goings-on involving Hannah's husband, Peter Crean-Smith, that have caused Hannah to become incarcerated at Gaze Castle in the first place, with Gerald Scottow her willing 'gaoler' (*TU* 112). A lack of coherence among the characters, the lack of a single centre of power in the novel, helps draw the reader to its electrifying constant: the agentive power of the surrounding landscape.

When Marian arrives from England, she is terrified of 'the rocks and the cliffs and the grotesque dolmen and ancient secret things' and 'appalled at the sudden quietness' (*TU* 15, 16). 'She was frightened now in an ordinary way, sick in her stomach, shy, tongue-tied, horribly aware of the onset of a new world' (*TU* 16), and yet as she contemplates the sublime view of the sea from Gaze Castle '[s]he looked and sighed, forgetting herself' (*TU* 17). The Scarren is under constant threat of further weather events, just like 'the night of the flood' (*TU* 9) when the fishermen's cottages and inn were destroyed and a car and its passengers were carried off the road and into the sea (*TU* 45). Gaze Castle's keeper Gerald tells Marian that the place was killed by a big storm some years ago. [...] It was quite a famous disaster, you might have read about it. And now the moor is just another piece of bog, and even the salmon have gone away. (*TU* 9)

For Marian's part, she 'had never seen a land so out of sympathy with man' (*TU* 11). The presence of the vast bog permeates the grand house.

The arrangements at Gaze contribute to the disorienting picture. It is unclear to Marian whether employer Hannah's mysterious incarceration is self-imposed (possibly through illness) or entirely beyond her control. At the end of an evening when the women have dressed and dined together, Marian is encouraged to take a turn in the garden alone. The gate appears to resist Marian's attempts to leave the grounds and, even as the gate gives way, she is spattered with earth and sand. Still, she feels involuntarily drawn back as the gateway appears to prevent her passing through. Murdoch portrays the resistance from Marian's surroundings and attributes an uncanny agency to the cliff-top lawn, as if the lawn itself is supervising Marian's movements:

> The garden was thick and magnetic behind her. Her desire to go out was gone. She was afraid to step outside. She stood paralysed in the gateway for some time, keeping her breathing quiet. The great lawn at the cliff top remained cold and attentive, visible yet unreal, waiting to see what she would do. (*TU* 54–55)

The garden's strange power paralyses Marian, its victim, while the lawn above enacts both passive observer and sentinel. She steadies herself, pulling her hands away from the stone archway. The apparent reversal of agential power at this juncture effects a sense of disorientation, contributing to the confusion and disquiet of ingénue Marian, and the reader. Marian's experience, under the scrutiny of the vertiginous landscape, draws attention to the omnipotence of Gaze Castle and its gardens.

The occasion provides an intense counterpoint to the idealized, seemingly carefree picture that the somewhat deluded Effingham Cooper, erstwhile student of the retired Max Lejour, associates with the landscape beyond Max's great house, Riders, as viewed across the valley from Gaze:

Beyond was the view of Riders, the black cliffs, the green islands, the windy sea, with near fishing boats and a steamer at the horizon. From a great height a silver aeroplane was coming down toward the airport. Effingham saw it all with a sort of shock. There was life, indifferent life, beautiful free life going forward. (*TU* 93)

The 'indifferent life' revealed by Effingham's picture of freedom comically betrays his complacent self-perception. Earlier on the train, en route to visit and read Greek with his old tutor, Effingham admires himself in a looking glass 'with an amused ironical affection. [...] He looked like a man; and he certainly passed, in the society which he frequented as a clever successful enviable one' (*TU* 67). He has known Hannah for four years and, aware of her predicament, seems tantalized by the fanciful idea that he might be 'the person with the most power, the only person who could really act' (*TU* 68). He imagines he might even have to endure an act of heroism in relation to 'the imprisoned lady' (*TU* 70), with whom he has become somewhat obsessed: 'No space-man about to step into his rocket was more meticulously fitted to go into orbit than Effingham at that moment was ready to fall in love with Hannah. He fell' (*TU* 71). Effingham takes little persuading to help Marian in her bid to set Hannah free, motivated presumably by the desire 'to gain possession of what he had so long worshipped' (*TU* 142). Murdoch is far from subtle in ensuring her reader does not take Effingham too seriously.

Murdoch's self-conscious acknowledgement of her engagement with the Gothic is embedded at *The Unicorn*'s centre. Marian and Effingham are about to attempt the (unsuccessful) abduction of Hannah in a bid to occasion the 'last day of the prison and the end of the legend' (*TU* 141). If *The Unicorn* were a piece of theatre, this is the moment when Murdoch would be said to break the fourth wall, explicitly acknowledging the contrivance of her Gaze Castle setting:

> It did at those moments seem possible that the sudden violence might produce, not the vanishing of a dream and the reassuring appearance of the ordinary world, but some shapes yet more Gothic and grotesque. (*TU* 141)

By naming the Gothic at this juncture, Murdoch appears to discharge her interior setting of its apparent artifice in an attempt to transgress perceptions of normality, freedom and love. After the botched abduction, Marian enters Hannah's 'lighted room from the darkness outside' (*TU* 153) and is greeted by 'a subdued cheerful murmur as if a decorous little party were in progress' (*TU* 153). The scene is a serene one, bathed in light, and Marian experiences a strange dissociative moment in the uncanny familiar:

> The golden group about her still seemed sheathed like seraphim from head to foot in serrated wings of light. Everyone seemed to have become very tall and elongated. She rubbed her aching eyes. Jamesie was saying something to her and giving her a cigarette and lighting her cigarette. She sipped the strong clean familiar whiskey. (*TU* 153)

The repetition of 'cigarette' emphasizes the dream-like confusion of her dissociation. Yet, the comfortingly familial scene also brings a strange relief to Marian that, viewed from the outside, persists as disquieting for the reader: 'I have been accepted into the family, that is what has happened. I have become part of the pattern' (*TU* 154). In a strikingly atmospheric reversal, Murdoch establishes an apparently soothing faux normality within Gaze Castle to juxtapose the ensuing exterior scene and its perilous setting.

With the story centred on the large Crean-Smith house, *The Unicorn* is often thought of as one of Murdoch's closed novels, akin to the twentieth-century 'crystalline novel' she characterizes as the 'small quasi-allegorical object portraying the human condition' in her essay 'Against Dryness', where she theorizes the novel and elucidates on her concerns about representation and realism (AD 291). *The Unicorn* is, however, the first of Murdoch's novels to truly inhabit (albeit largely constrained to a single chapter) a seemingly boundless sublime external reality: an outer world populated with the agentive forces of the more-than-human. To experience the sublime is to experience 'the vastness of nature' (SBR 263), asserts Murdoch.[7] In her explication of how such a picture might emerge,

---

[7] It should be noted that, later, in 'The Sublime and the Beautiful Revisited', Murdoch prioritizes the 'manifold of humanity' over 'physical nature' (SBR 282) as the more challenging picture of the sublime.

she deviates from the imperious and rational Kantian-man idea, to connect the Murdochian perceiver of sublimity with an 'undramatic, because un-self-centred, agnosticism which goes with tolerance', which she ultimately redefines as 'love' (SBR 283).[8] In fact, Murdoch connects her conception of the sublime to freedom: '[f]reedom is knowing and understanding and respecting things quite other than ourselves' (SBR 284). Murdoch puts her theorizing to work in this pivotal chapter of *The Unicorn*, wherein the action moves from the enclosed Gothic interior to picture a contingent real-world transcendent reality wherein the menace continues to feel undeniably present.[9]

Effingham is lost. He has discovered how Marian's plan to remove Hannah from Gaze has come to be thwarted, and he is last seen by Alice heading in the direction of the dangerous peat bog in a self-absorbing fit of rage. Murdoch describes how the 'accusing remorseful thoughts buzzed in his mind, making him blind and deaf' (*TU* 159) to his surroundings. Eventually, calming down, he begins 'to apprehend what was outside him, and became gradually aware that he was surrounded by a vast silence' (*TU* 159).[10] After the internalized frenzy, he is suddenly alert to the bleak landscape and realizes he is, indeed, lost. He becomes unnerved by 'the darkness, the emptiness, the absence of human activity' (*TU* 161). Effingham mistakes an object standing sentry up ahead for a familiar dolmen. In fact, it is a tree. 'Trees were individuals in that part of the world' and, in finding unexpected affinity in it being 'upright like himself'(*TU* 161), he is reluctant to leave it behind. 'The tree was at least a sort of

---

[8] 'Kant abolished God and made man God in His stead. We are still living in the age of the Kantian man, or Kantian man-god' (SGC 365). He is 'attractive' but 'misleading', says Murdoch. 'He is the offspring of the age of science, confidently rational and yet increasingly aware of his alienation from the material universe which his discoveries reveal; and since he is not a Hegelian (Kant, not Hegel, has provided Western ethics with its dominating image) his alienation is without cure. He is the ideal citizen of the liberal state, a warning held up to tyrants. He has the virtue which the age requires and admires, courage' (SGC 365–66).

[9] 'The Sublime and the Beautiful Revisited' (SBR) was first published in the *Yale Review* in December 1959, four years before Murdoch published *The Unicorn*.

[10] The ambient sound of silence that engulfs Effingham contributes to the animistic nature of this novel. To this effect, Dooley remarks that the novels generally 'are full of silence of all kinds—often not just the absence of sound, but an almost palpable quality in itself. Her silences come in different qualities, moods and textures: they can be terrible, annihilating, brooding, horrible, intense, idiotic, accusing, sinister, profound, uncanny, fox-like, alarming, ominous, menacing, vast, immense, merciful, even happy' (Dooley 2022, 107).

shelter, a sort of house, a sort of *place* with some kind of significance. Now all around him there was nothingness and nowhere' (*TU* 162), we are told. In the absence of others like himself, he demonstrates capacity to see personhood in the tree. At this moment, his vulnerability is aligned with a total dependence on, or a submission to, a landscape existing in and of itself.

The landscape is animate: 'Nothing human lived up here' (*TU* 162). Effingham 'knew perfectly well that there were no such things as fairies or spirits or malevolent non-human agencies' (*TU* 163), yet the stars produce a 'semblance of diffused light' which is surpassed by the mysterious green illumination on the ground with its 'intense hard brilliance', encircling him in 'a fainter line' (*TU* 163). Fearing the sinister and the supernatural, he attempts to rationalize the phenomenon of the flares of 'fairy fire' he can see with explanations of 'chemical causes and constituents' (*TU* 163).[11] In fact, the presence and behaviour of the strange flares help him to realize that he has been walking in circles. Effingham is suddenly aware of his human vulnerability. He sees himself as an intruder in the landscape. He has become aware of the striking opposition of profound darkness created by the radiating bioluminescent light and conceives of 'menace round about him of presences to whom human things were abhorrent' (*TU* 164). Recalling tales of 'morasses […] that would engulf a man' (*TU* 165), Effingham realizes that he is indeed stuck in the bog and sinking quickly.

At this moment, Effingham is driven to submit to the landscape as Randall had submitted himself among the lines of roses at the nursery. However, Effingham's moment of solitary epiphany in nature proves considerably more dramatic. He must face the perilous forces of the quagmire and confront death. As he sinks deeper into the mud, he senses the drifting away of self and begins to apprehend the significance of the surrounding world. He experiences a certain clarity and an overwhelming feeling of love:

> What was left was everything else, all that was not himself, that object which he had never before seen and upon which he now gazed with the

---

[11] Fairy fire is also known as fox fire (derived from faux fire) or bioluminescence.

passion of a lover. And indeed he could always have known this for the fact of death stretches the length of life. Since he was mortal he was nothing and since he was nothing all that was not himself was filled to the brim with being and it was from this that the light streamed. This then was love, to look and look until one exists no more, *this* was the love which was the same as death. He looked, and knew with a clarity which was one with the increasing light, that with the death of self the world becomes quite automatically the object of a perfect love. (*TU* 167)

Like Randall and Rainborough before him, Effingham is presented as a picture of solitary unselfed love in nature. On this occasion, however, the scene is configured in the bleakest of natural environments. As before, Murdoch depicts the occasion of a perfect vision of love outside of self as fleeting.

In Effingham's case, the 'depersonalized, abandoned [...] self' (*TU* 167) returns with 'a frenetic desire to live' (*TU* 168). It is not long before Effingham is rescued by Denis and returned to Gaze Castle. Those around him think he has completely taken leave of his senses as, glass of whiskey in hand, he evangelizes:

> Before the self vanishes nothing really is, and that's how it is most of the time. But as soon as the self vanishes everything is, and becomes automatically the object of love. Love holds the world together, and if we could forget ourselves everything in the world would fly into a perfect harmony, and when we see beautiful things that is what they remind us of. (*TU* 173)

Again, Murdoch constructs epiphanic experience as selfless love revealed to the solitary individual in landscape, in this case, in a precarious agentive landscape. But the moment has quickly vanished. Effingham's thoughts begin to disappear too, lingering on only as empty words as he converses with the group back at Gaze and contemplates their 'spiritual amity', configured 'by their joint concern for him' (*TU* 173). Murdoch continues, 'what a perfect object of love they made, they-loving-him, together. So love, making an unchecked circuit, returned to himself' (*TU* 173). With this last line, Murdoch wryly confirms the inevitable return of the 'fat relentless ego' (OGG 342) of Effingham's recently vanished old self.

There remains still much to play out in the human drama of *The Unicorn* but, at the end of the novel, Marian, already deeply shocked by Hannah's death by suicide, is appalled to see the devastation to the valley between Gaze and Riders after a night of torrential rain sends 'a huge torrent, wide, brown and turgid, roaring down the centre, dividing the two houses' (*UN* 246). The 'bog had released its waters' (*UN* 246). The textual repetition of flood that book-ends the novel, reconfiguring the landscape before the start of the novel and whose effects are yet more devastating at the end, is an inherent part of the circularity of Murdoch's craft acknowledged by many critics, including Nicol (2004, 57) and Horner. Horner notes, for example, the story of the unnamed man who could not be reached before he died in the bog two years before the novel begins, which sets up Effingham's predicament and readers' expectations of his fate. His unexpected recovery, in turn, raises the (false) hope for Hannah's own moment of revelation and survival (see Horner 2010, 78). The devastating landslide brings events at Gaze to a natural close.

The Australian ecofeminist and self-styled 'philosophical animist' Val Plumwood asserts that animism demands the telling of the 'self-inventive and self-elaborative capacity of nature, about the intentionality of the non-human world' (2014, 447, 449). She argues that 'any rediscovery of "tongues in trees" (*As You Like It*, II.1.563) is a matter of being open to experiences of natures as powerful, agentic and creative, making spaces in our cultures for an animating sensibility and vocabulary' (Plumwood 2014, 451). *The Unicorn* reveals Murdoch experimenting in new and compelling ways with her representations of non-human life that begin to take account of the agentive qualities of the natural world. These creative decisions are informed by her acknowledgement of the significance, indeed the equity, of these other lives.

Murdoch's equitable view of lives stems from her profound love of the world, and her search for the 'good in everything', to echo Duke Senior (*As You Like It*, II.1.564). Tony Milligan recognizes her love as 'reaching out beyond love for other humans' and therefore inclusive:

As a simple point about what Murdoch is textually committed to, love of a legitimate good sort is presumed to extend beyond persons and also beyond creatures. It extends to love for non-human objects and artefacts; and even to love for the planet and/or nature itself. (Milligan 2022, 468, 470)

He cites Murdoch in support of this recurring theme: 'Human beings love each other, in sex, in friendship, and love and cherish other beings, humans, animals, plants, stones' (*MGM* 497). Although for Milligan, at least, some of Murdoch's loves seem 'a little unlikely, even absurd' (2022, 470). Stones belong in this category as far as he is concerned. Yet the idea of attending to and loving even stones is a subject of discussion in Murdoch's philosophy, and is also a common feature of so many of her novels.

Milligan proposes that love (rather than the Good) should be regarded as Murdoch's sovereign concept and asserts a 'dual […] conception of love' which, in his view, assists in assuming responsibility for the care of the environment: 'love as a special emotion' and 'love as eros' (2022, 472). He argues the first as a metaphoric replication of the second, explaining that while 'we cannot love discrete future persons in the sense that requires the possibility of companionship, we can nevertheless *love* future generations, or humanity itself, or the planetary ecosystem, or any number of less tangible objects' (Milligan 2022, 472). Of course, experience of the second (love as eros) can assist in encounters that engender the first (love as a special emotion): a deep love of a particular rock or tree, for example, more likely than not engenders a love for all rocks or all trees.

Murdoch acknowledges the human propensity for love in responding to the natural world in her essay 'The Sovereignty of Good over other Concepts'. Instinctual love for Murdoch, as presented here, is the embodied sense of attention and emotion engendered by an enchanting world:

> It is so patently a good thing to take delight in flowers and animals that people who bring home potted plants and watch kestrels might even be surprised at the notion that these things have anything to do with virtue. The surprise is a product of the fact that […] beauty is the only spiritual thing which we love by instinct. (SGC 370)

And she is able to find intrinsic value in its beauty, in the separate existence of other things, other lives. In her essay, she emphasizes the primacy of the sub-conscious, immersive or unexpected appreciation of the natural environment. When she characterizes the existence of 'animals, birds, stones, trees' as 'alien, pointless' and 'independent' (SGC 369–70), it is not to devalue these lives: they are strange, they are 'pointless' in that they are not there to serve a direct purpose to human lives, and their existence functions quite separately to us. Murdoch expresses their intrinsic value, her purpose to engender humility in the gaze of the self-regarding sovereign individual, to prioritize equity and, therefore, justice. Murdoch paraphrases Wittgenstein: 'Not how the world is, but that it is, is the mystical' (SGC 370). The natural world should elicit our wonderment, value and care. The aesthetically pleasing cultivated landscapes of her earlier, formally tighter novels are evolving into the more expansive, animistic and sublime landscapes of some of her later larger ones.

Murdoch's preoccupation with lithic matter—pebbles, rocks, stones, boulders, mountains—intrigues many of her readers. Rocks and stones feature in many of the novels and, much like her character Charles Arrowby, Murdoch was a compulsive and prolific collector of stones herself. To Brophy she writes on 14 April 1963, 'I picked up a number of *stones* [at Charmouth] and will bring you one' (*LoP* 243). In his memoir, Bayley recalls countless stones submerged by other clutter in Murdoch's study at Cedar Lodge, Steeple Aston:

> It grieved me then, and still does, that these stones, once so naturally clean and beautiful from continual lustration in a stream or by the tides of the seashore, should have become as dusty and dead-looking as everything else in the house. But this never seemed to bother Iris in the slightest. The stones for her were Platonic objects, living in some absolute world of Forms, untouched by their contingent existence as a part of the actual and very grubby still life that surrounded us'. (1999, 152)

Anecdotally, there are said to remain hundreds of stones which Murdoch collected over the years in the garden at Charlbury Road, Oxford. In *The Sea, The Sea*, Charles obsessively gathers 'pretty stones' (*TSTS* 11) into a rocky trough at Shruff End:

> I cannot stop collecting stones and the trough is overflowing even though I have put some of the best ones into my lawn border. […] It is a good way to display the stones, but will the earth discolour their undersides? (*TSTS* 74–75)

In *The Nice and the Good*, twins Henrietta and Edward Biranne collect stones, arranging them decoratively at Trescombe House: 'There's something special about *every* stone' (*NG* 12), claims Edward. In *The Green Knight*, Moy proves able to move stones telepathically, by affording them an intense form of attention:

> She was gazing at one stone in particular, golden brown, shapeless as crushed brown paper. She moved reaching out her hand towards it. After a moment the stone shifted slightly, it rocked then slid evenly forward off the shelf and through the air into her open hand. (*GK* 20)

Moy thinks her ability unusual: 'She accepted it as a strange not unfriendly presence or form of being which joined her life with the life of things. Only sometimes, for it had various manifestations, it frightened her' (*GK* 20). Towards the end of the novel, Moy notices that the 'curious powers which had once alarmed her had now been withdrawn' (*GK* 461). The stones 'lay now inert, her things, no longer related to her by mysterious ties' (*GK* 461). She feels an intensely deadening feeling from 'the loss of contact with innumerable entities whose relationship with her she had taken for granted' (*GK* 461). Murdoch seems to connect her erstwhile telepathic skills to an innocent and unconstrained childhood. In a nod to the posthuman, Murdoch has Moy return the conical stone to where she feels it might belong:

> She walked down the beach to a line of low rocks and took the heavy conical stone out of its bag. Then climbing up a little, she placed it on top of a flat rock. Perhaps the stones could signal to each other? But could she interpret the signal? She climbed down and walked back and wandered about near the shore upon the grass between the stones and the hills but felt nothing and saw nothing. She wondered whether she should leave the stone here upon the rock, where it was rather conspicuous. Perhaps someone else would find it and take it home. Could that be a good thing or a bad thing? Or should she put it into the sea? Would it like the sea? It was

not a sea stone. Yet, in hundreds and thousands of years it would become a sea stone, the runes would be washed away, its sharp cone would be softened into a hump and sea creatures would live upon it. She started to climb up the rock again to retrieve the stone, then decided not to. What did it matter? It was just a stone. It was nothing. She was nothing. (*GK* 463)

This brief sequence at the shoreline, on 'the day after Clement had shown Louise the sea and the beauty of the world' (*GK* 462), explores what it means to sense place and lithic belonging. Moy conceives of the vastness of space and time that brings about change in a stone, a slow-burn gradual change that will continue long after she is gone from the world. Everything (and nothing) matters: 'It was just a stone. It was nothing. She was nothing' (*GK* 463). Murdoch conveys a sense of our time on Earth as being brief, transient and gloomily inconsequential.

In all of these examples, the unique qualities of each stone, observable through close attention, is valued and celebrated. In each interaction, a character's regard for the inherent value of each stone is its overarching message. Such interactions with lithic matter reciprocally help to reveal some of the innate qualities and deeply held beliefs of their human counterparts. The co-dependence of human character and lithic matter is fundamental to this conception. Murdoch's 'vision of humanity as part of a living, palpable planet' (Rowe 2019, 112) is inherent to these scenes and aligns with cultural ecologist and philosopher David Abram's conception of a 'sensuous world' (1996, ix). Abram's discernment of an 'age-old reciprocity with the many-voiced landscape' coheres, for example, with Murdoch's picture of Rainborough's 'outer world [...] mingled with thought and transformed into an inner substance' (*FFE* 121). This is an early iteration of the sort of moment that is to be understood as the novelist's awareness of the sensuality of an animate world. That is to say, Murdoch's conception clearly correlates with Abram's own apprehension of animism.

For Abram, Earth's matter is innately vibrant and dynamic: 'To the animistic sensibility *everything* is animate—*everything moves*—although some things, like granite boulders, move much slower than other things, like [...] swooping hawks' (1996, 278). Stone, ancient and unyielding, lends a distinctly different temporal perspective to the human story. It is 'primal matter, inhuman in its duration', argues Jeffrey Cohen (2015, 2). He asserts that '[stone's] injunction is always to step out of the breathless rapidity of anthropocentric frames and touch a world possessed of long futurity and deep past' (Cohen 2015, 33). He notes the communications that have been relayed down through the ages from prehistoric times through the power of stone: 'a stone-etched countervision invites reflection on what it means to inhabit a world that is at once potentially indifferent to humanity yet perilously continuous' (Cohen 2015, 60).[12] Such ideas force us to set our own human history into the broader context of vaster planetary scales and, indeed, take a relativized view of human (in)significance. This is precisely what Murdoch appears to address with Moy and the stones at the end of *The Green Knight*.

The Murdochian Unself

The central feature of Murdoch's emerging animistic sensibility is her rejection of the hierarchy of primal subject that subjugates passive object. Instead she projects an awareness of the co-dependence of all life. According to Julia Jordan, in Murdoch's novels,

> there are those who get along with objects and those who don't: those who like the stone for its stoniness, and those who try to appropriate it, to read it; those who co-opt its stoniness and try to make it signify. (2012, 365)

One infers, from these remarks, Murdoch's broader concern about a general lack of care for things. Jordan explains that 'Murdoch's work explores and ceaselessly interrogates this attitude, and its ethical and ontological legitimacy, by contesting conventional hierarchies of subjects and objects' (2012, 365). By bringing these non-human elements of her work to the

---

[12] Cohen names sites such as Avebury as '[d]evices for the lasting conveyance of story', where 'such architectures communicate long after their human co-dwellers vanish' (2015, 60); Murdoch visited Avebury as a schoolgirl in 1938 and wrote an essay, 'Millionaires and Megaliths', for *Badminton School Magazine* about the respect being shown for the Druids in the reinstatement of their ancient stone circle, republished in *IMR*10 (2019, 1–3).

fore in her novels, she is rejecting anthropocentrism in favour of a more equitable world. In the meantime, Murdoch's more honourable characters tend to 'exercise a recognition and appreciation of the alterity of things, of their ontological integrity and separateness' so that 'an awareness of alterity comes to constitute a way of being in the world, an ethical mode that willingly accommodates difference and is capable of surprise at the very "thinginess" of things' (Jordan 2012, 365).[13] Murdoch's depiction of the non-human world brings context to any fixation on human affairs.

Murdoch is recognized by environmental ethicist Mick Smith for reaching beyond anthropological concerns. Smith rejects the human exceptionalism that claims Earth's finite resources as humankind's to exploit, and he calls for a significant shift in the relationship dynamic to take account of the co-reliant interest and shared potential of an animate Earth. He identifies tools in Murdoch's early philosophical writing which contest the sovereign principle of human dominion over planet Earth. In contemplating the natural world, Murdoch begins by renouncing the self-obsession of the sovereign individual, expressed in these important words: 'A self-directed enjoyment of nature seems to me to be something forced' (SGC 369–70). Murdoch highlights the instinctual love, 'the self-forgetful pleasure', brought about by being immersed in natural surroundings (SGC 369). The '"self-directed enjoyment"' (SGC 369) that Murdoch rejects is, for Smith, also a rejection of the 'idealized human being' exploiting Earth for his or her purposes (2011, 38–39). The 'self-forgetful' encounter (SGC 369) with nature is the route to unselfing for Murdoch, a concept inspired by Simone Weil's *decréation*.[14] In *Gravity and Grace*, Weil regards herself as the 'unwelcome third' in a relationship with the Earth and with God and, by quoting from Racine's *Phèdre*, she appears to suggest that perfection can only be achieved once she herself is no longer part of the world:

---

[13] On receiving a gift from the artist Harry Weinberger, Murdoch remarked, 'how nice objects are—I'm glad we live in a thingy world' (*IMAL* 588).

[14] Silvia Caprioglio Panizza explains that 'although Weil's religious attention has an ethical dimension, and although Murdoch's secular attention has religious inspirations, the large overlapping between their concepts of attention also contains differences'; for a comprehensive assessment of Murdoch's debt to Weil, and the similarities and differences between Weil's '*decréation*' and Murdoch's 'unselfing', see Caprioglio Panizza (2022, 19).

> *Et la mort à mes yeux dérobant la clarté*
> *Rend au jour qu'ils souillaient toute sa pureté.*
> [And Death, robbing my eyes of their light,
> Restores to the day they sullied all its purity.] (1952, 36)[15]

Phèdre is dying. Debased by her own deceit, she has admitted her responsibility for the death of her stepson Hippolytus to Theseus, and has poisoned herself. Weil adopts this tragically self-annihilating metaphor to express her own conviction that it is only by removing herself that perfection in the world is truly attainable. Murdoch rejects any notion of self-annihilation by advocating presence and clear vision in a contingent and uncertain world: to see the world as it really is, and not in the consolatory form that we might attempt to envisage. Rather than removing self, as Weil proposes, Murdoch's method is to aspire to, and work towards, a state of selflessness (unselfing) in order to attend to the other and to sense the immanence of the world.

Murdoch acknowledges the value of being in and attending to nature when it allows us to 'clear our minds of selfish care' (SGC 369). Moreover, Murdoch's 'hovering kestrel' (SGC 369) provides sufficient evidence for Smith to insist that Murdoch's encounters with the natural world should not be read as indulgent self-acclamation when it might appear that she directs herself inward towards her own concerns rather than outward towards the more-than-human lives she observes. Murdoch's act of attention to the other is critical here. Through this act of unselfing in momentary contemplation of free, independent more-than-human life, she is clearly resisting 'the tendency to recuperate and resituate her thoughts within those self-reverential frames that reduce all else (including kestrel) to beings of merely instrumental value' (Smith 2011, 39). In this way, the Murdochian unself provides the bedrock for an environmental ethics. When we attempt to attend selflessly to 'how the world is', the 'sheer alien pointless independent existence' of beings acknowledges presence and a materiality that exists freely from us and separately in its own right (SGC 369). Murdoch wants us to attend to the world with a just and

---

[15] In *La pesanteur et la grâce* (1947), Weil (or her editor) appears to have misquoted Racine when she writes: '*Et la mort à mes yeux ravissant la clarté*'; Racine's Phèdre says: '*Et la mort à mes yeux dérobant la clarté*' (1992, V.vii.1641).

loving gaze, to contemplate nature's form because she believes such perfection hints at what good is, and therein lies its ethical value.

Murdoch only ever presents the hard work of attending to an imperfect world as bold ambition, but these attempts at humility give us the opportunity to notice the significance of the more-than-human that is at the mercy of humanity's long-established sovereign power. Silvia Caprioglio Panizza reminds us about the challenges of attempting to administer a 'non-anthropocentric perspective' to the world, 'given [our] humanity and sociohistorical context' (2022, 9). However, she thinks now is the 'time to at least ditch the *assumption* that moral thinking is properly or primarily about humans' (2022, 9), an assumption she believes Murdoch made. Caprioglio Panizza nonetheless asserts that Murdochian philosophy offers tools 'that are among the best for moral thinking generally' and seems able to go along with Smith in his suggestion that through Murdoch we can re-think 'human dominance over nature' (2022, 9). Smith argues keenly for a Murdochian ethics that is not limited to anthropological concerns, yet he insists that this evident lead on environmental ethics is an 'unintentional' one on Murdoch's part (2011, 50). Caprioglio Panizza is inclined to agree with Smith here. If by 'unintentional' Smith means that Murdoch creates images of the natural world merely to illustrate her ethics of attention, then I think it is difficult to agree with him on this point. It is true that Murdoch does not offer specific tools relating to animal ethics, for example, but Caprioglio Panizza is right to suggest that 'a better starting point' should be 'Murdoch's emphasis on love, vision and particularity': 'by returning to the individual, reality, and felt experience, Murdoch's ethics makes it easier both to approach animals, and to think of ourselves as animals' (2022, 9). It is in this state of being, immersively a part of nature, that I think elements in Murdoch's fiction, her journals and her poetry present powerful evidence that supports the counterargument—that her lead on environmental ethics is a considered one. One place where these ideas coalesce is in *Nuns and Soldiers*, where the powerful agency of lithic matter on an altogether vaster and more imposing scale, the magnificent limestone Alpilles mountain range, intervenes in the young artist Tim Reede's moral development.[16]

Two sequences of *Nuns and Soldiers* are set in the Provençal region of France, a location where Murdoch and Bayley regularly spent time with their friends, Natasha and Stephen Spender. Inspired by her visits to this region, Murdoch placed key moments of the French sequences of *Nuns and Soldiers* in settings created through a 'collage of several far-flung focal points of her daily wanderings' (Spender 2019, 67). On these visits, returning from walks in the surrounding landscape, Murdoch would work alongside her host in the garden 'in a calm almost Buddhist atmosphere of contemplation', records Natasha Spender (2019, 67) in an essay written in 1988 and eventually published in a volume to mark Murdoch's centenary in 2019. 'One caught her compassionate attitude' to small creatures 'like an infection', Spender continues, 'however minute and unremarkable the creature; one could even come to love green-fly or red spiders' (2019, 66). In *The Philosopher's Pupil*, the Murdoch novel that follows *Nuns and Soldiers*, Gabriel is the character that gives creative expression to this tenderness towards small creatures (and things):

> One of the qualities of her interior castle she had acquired from [her son] Adam—a sort of animism whereby everything, not only the flies which had to be caught and let out of windows, the wood lice which had to be tenderly liberated into the garden, the spiders which were to be respected in their corners, but also the knives and forks and spoons and cups and plates and jugs, and shoes, and poor socks that had no partners, and buttons which might become uncherished and lost, had all a life and being of their own, and friendliness and rights. (*PP* 61)

From the compassion for small creatures to the grandeur of the surrounding landscape brought into play in *Nuns and Soldiers*, Spender's essay records the seismic shift in Murdoch's source material and her imaginings.

Spender celebrates the 'strangely beautiful land' of Les Alpilles in her essay (2019, 66). Les Alpilles comprise a low-level chain of limestone

---

[16] A section of this chapter has been developed from an essay first published in a special edition of *Ebc—Iris Murdoch and the Ethical Imagination: Legacies and Innovations* (see Oulton 2020).

mountains that lie twelve and a half miles north of Avignon and stretch from Mont Ventoux in the east to the Parc National des Cévennes in the west. Spender senses the lithic power of the 'east-west chain of limestone mountains, small in size yet dramatic in scale, which dominates the alluvial plain bordering the Mediterranean Sea', where she feels

> possessed by its mystery and by the sense it gives one of antique gods—the upward surge of the limestone suggesting the overpowering gesture of some subterranean Pluto hurling them heavenwards. In this little outpost of the great Roman Empire 'in hither Gaul' one can scarcely help but feel an echo of the Roman sense of the implacable power of natural forces and the modest brevity of life for any human inhabitant. (Spender 2019, 65–66)[17]

The articulation of Spender's own sensory awareness of the majestic landscape, of sentient forces and temporal sublimity, aligns closely with the early pagans whose animist beliefs are said to have all but been destroyed by the arrival of Christianity just over two thousand years ago.

In a ground-breaking, mid-twentieth-century essay, Lynn White identifies humanity's predicament. He is convinced that human ecology is 'deeply conditioned by beliefs about our nature and destiny—that is, by religion' and by the innate obsession of Westerners for ever-greater progress (which persists today) and which, according to White, is rooted in Judeo-Christian principles (1967, 1205). Western Christianity, he claims, is 'the most anthropocentric religion the world has seen', when 'man is made in God's image' (White 1967, 1205) and its teachings imply that the world is ripe for exploitation. By stark contrast, he argues, '[i]n Antiquity every tree, every spring, every stream, every hill had its *genius loci*, its guardian spirit', and as such engendered respect and compassion (1967, 1205). White asserts that, in 'destroying pagan animism, Christianity made it possible to exploit nature in a mood of indifference to the feelings of natural objects' (1967, 1205) and create the now centuries-old view of human dominion over nature, the same view that Smith decries half a century later. Attempts to counter the man-versus-nature binary here, however, risk appearing supplanted by a religious

---

[17] Spender's reference to 'hither Gaul' (Gallia Citerior) reaches back evocatively beyond the Roman Empire to a western Europe of the Iron Age period.

dualism. The nuance of White's important essay risks getting lost in this brief discussion of it, not least his endorsement of St Francis of Assisi, the 'greatest spiritual revolutionary in Western history' despite the failed attempt to champion the 'equality of all creatures' (1967, 1207). What remains incontrovertible, however, is White's avowed first precept, that '[w]hat we do about ecology depends on our ideas of the man-nature relationship' (1967, 1206). Murdoch's own ideas on human and more-than-human relationships (to use today's more appropriately equitable terminology) are found not only in her philosophical thinking but also in her fiction.

Murdoch expands the temporal perspective of *Nuns and Soldiers* when she weaves the ancient craggy limestone of Les Alpilles, the radiant pulsating pool and the fast-flowing canal (inspired by the canal de Craponne) into the actions of her superficial self-believer Tim Reede. Such temporal reach engenders meaning in the rocky landscape. Remove the animistic quality of these material elements, ignore the landscape's lithic and fluid power, and Tim's story loses meaning. Cohen contends that 'thinking beyond anthropocentricity' insists on a reset of temporal perception (2015, 9). Bennett avers that humans tend to orientate themselves by interpreting the world 'reductively as a series of fixed objects' (2010, 58). Yet with rock, for example, we are dealing with 'mobile, internally heterogeneous materials whose rate of speed and pace of change are [merely] *slow* compared to the duration and velocity of the human bodies participating in and perceiving them' (Bennett 2010, 58). Rock is, then, a site of human and non-human interaction. Spender's essay provides prescient and eloquent expression of Serenella Iovino and Serpil Oppermann's 'storied matter' (2014, 1), producing the signifying interplay of more-than-human forces over time. Murdoch's depiction of Tim's embodied interactions with the rocks and the canal is immersive in both literal and metaphorical senses so that when, through his sensuous body, Tim submits to the 'Great Face' (*NS* 159) and the watery features of the landscape, Murdoch is inviting her reader to interrogate the frailty of human experience in relation to the unyielding, stoney presence of the physical environment.

When Tim arrives at Gertrude's house in France, he sees human order in the poplars 'set in rows with clean light paper-brown trunks and high twinkle of leaves' (*NS* 148) across the stream. Where nature has encroached on an abandoned olive grove, the trees appear to fall about 'in grotesque attitudes, splitting into huge semi-recumbent forms possessed of elongated faces and writhing bodies' (*NS* 148). The imposing skyline replete with umbrella pines and magnificent rocks dominates the scene and yet, asserting a sense of his human sovereignty as the 'monarch of all he surveyed' (*NS* 149), Tim sees only uninhabited landscape.[18] Gertrude has commissioned a drawing of her house from him, but Tim finds himself both obsessed with and intimidated by the surrounding rocky landscape: what he sees, he is quite unable to draw. Murdoch believes that all that we are able to notice, set value to and love constitutes progress in the moral life. She maintains that 'what we literally see is important. Perception is both evaluation and inspiration, even at the level of "just seeing"' (*MGM* 329). For this reason, she goes to some length in *Nuns and Soldiers* to express Tim's artistic potential sympathetically, almost affectionately, and to explain that it is his idleness and lack of confidence that fail him. We perceive through all of our senses, and Murdoch wants her reader to recognize Tim's talent as instinctual, even primal, when much of what he expresses in his more successful artistic work is just that: learned through the sensing body. Murdoch writes that 'dark knowings were effective in Tim's mind. [...] He had a "feeling" about plants, [...] he instinctively understood how feathers had to grow to make a wing', and '[h]is body told him about gravity, about weight, what falling was, what flowing was' (*NS* 127). Murdoch expresses the (often unspoken) naturally human propensity to perceive the world through the body which not only resonates with Abram's 'sensuous world', but is in consonance with Maurice Merleau-Ponty's work on the world inhabited by the body which I discuss later on in this book.

In *Nuns and Soldiers*, Tim explores his new surroundings close to twilight when he chances upon a vast intimidating rock that he sees anthropomorphically as a 'great face', and beneath it a rock pool that Tim can 'scarcely believe [...] to be a work of nature'—a 'large circular pool of very clear water' (*NS* 159, 156). Dusk causes him to retreat into the

---

[18] This image recalls Murdoch's quarrel with Sartrean existentialism in her essay 'Against Dryness': 'We are not isolated free choosers, monarchs of all we survey, but benighted creatures sunk in a reality whose nature we are constantly and overwhelmingly tempted to deform by fantasy' (AD 293).

3 The Emerging Animism in the Novels: *The Flight...*

house but, drawn again to the stone basin the next morning, he sees by day it is 'clearly a work of nature' but 'no less awe-inspiring' (*NS* 157). He is startled by the pool's radiant beauty which 'simply quivered in perpetual occult donation and as perpetual renewal' (*NS* 158). His heart is 'so filled with joy' that his instinctual embodied reaction is to 'clutch at it with both hands' (*NS* 158). As cognitive response comes into play, he decides he will not bathe in the pool for fear of polluting the pure water or 'interrupt[ing] its sibylline vibration', although he does allow himself to 'break the surface with his fingers' and feel the chill of pure radiant water (*NS* 158). Unpacking his art materials, he gets to work: 'He had that pure clean blessed beginning-again feeling. He was full of grace. He sat down, completely happy, and began to draw' (*NS* 158).[19] Here, Murdoch captures in her descriptions what Bennett calls '[*t*]*hing-power*: the curious ability of inanimate things to act, to produce effects, dramatic and subtle' (2010, 6). Bennett advocates a 'vitality of matter' that constitutes the expression of the animism—described by Abram, by Lynn White before him and experienced by Spender—that contributes the 'agentic force' (2010, x) which Tim perceives affectively through his body in the numinous presence of the radiant pool and the mountainous landscape.

Tim, then, becomes artistically productive among the mighty rock formations when, in an expression of lithic agency, the 'subject somehow took charge of him and conveyed some of its grandeur into his vision' (*NS* 159). Later, climbing up over the 'rocky skyline' and gazing down 'into another land', Tim has lost his imperious demeanour and instead senses pure pleasure at the sight of a 'flashing river' and the 'joyful commotion of its copious waters' (*NS* 161). A brief swim in the 'headlong

---

[19] Rebecca Moden asserts that the painter Harry Weinberger's 'identity hovers behind the character Tim Reede, […] who like Weinberger is captivated by the enigmatic Alpilles mountain range, its colours constantly fluctuating resisting interpretation'; Moden connects this painterly obsession to the work of 'philosophers, neuroscientists, linguists, psychologist and computer scientists', explaining how they 'have joined forces in their quest to understand the nature of colour vision', and here she cites colour theorist Evan Thompson, to elucidate the growing recognition that 'colour lies at the intersection of mind and matter, perception and the world, metaphysics and epistemology'. Examining landscape in this way, Moden avers that '*Nuns and Soldiers* invites an ecocritical interpretation which reinforces its relevance to contemporary philosophical debate regarding colour'; it suggests that Moden, too, senses the immersive effects of Murdoch's rendering of Les Alpilles, (Moden 2023, 191–92, cites Thompson 1995, 2).

force of grey curling water' shocks him 'with cold fear' when he realizes that the water ahead disappears downhill 'into a white chaos' and then enters the mouth of a subterranean tunnel (*NS* 162).²⁰ Tim submits to the overwhelming power of the mountain chain's materiality that momentarily makes him acutely aware of the 'fragile mortality of his own body, which comes to most of us as rare reminders, too soon forgotten' (*NS* 161). However, Tim's tendency towards duplicity is set to complicate his talent. He deceives himself, as he will also later deceive Gertrude.

Tim's communion with nature is not restricted to his time spent in Provence. Rather, later in the novel, on the day he decides to leave Daisy for good and is asking himself how he could possibly have thought of leaving Gertrude, Tim undergoes a moment of spiritual enlightenment in London's Hyde Park:

> He could see the trees, the huge quiet planes, with their immense friendly peeling trunks and the vast dangling swing of their downward-reaching branches covered with feathery leaves. He walked on over the grass which was dry and warm and bleached to a faded gold, and it made a soft springy sound under his feet. He could see in the distance the line of the lake and the Serpentine bridge. Then suddenly his knees gave way, he knelt down and lay prone upon the grass. Like an orgasm, like a birth, something wrenched his body and then left it, leaving him utterly limp. A warm wave had broken over him and now flowed on and on. A wave of pure thoughtless happiness which made him, with his face in the dry grass, moan and moan with joy. (*NS* 397)

Following this private moment of ecstasy in the most public of places, Tim needs very little encouragement to return to Provence in pursuit of Gertrude.

---

[20] Bayley recalls that 'In the great heat of July we used to plunge into the ice-cold "agricultural", an old irrigation canal that wound through the steep contours of the hills, running swiftly among the dense thickets of green canes and rosemary and cypress that bordered abandoned apricot and olive groves. They seemed to have reverted to a wild state. [...] A gripping sequence in Iris's novel *Nuns and Soldiers* was inspired by our discovery of a tunnel in the *maquis*-covered hillside. We could see light at the end of it and ventured to wade through. The hero of the novel had a more exciting adventure in a subterranean stream. But the magic place, the overpowering heat at midday, and the grey alpine water rushing on its mysterious course through the abandoned country—these were just as Iris described them' (1999, 218).

The encouragement arrives in the form of a letter from Veronica Mount—to which he reacts impulsively—'Gertrude still loves you, needs you and wants you to come back', Mrs Mount has written; 'She has not said this, but I believe it to be so. She is at present at the house in France, alone as far as I know' (*NS* 413). Tim arrives at Gertrude's French retreat, *Les Grands Saules*, in the 'gathering dark' (*NS* 419), and from the terrace can see through the window that Gertrude is not only not alone, but sitting across the table from the Count, holding hands. For Tim, the moment instantly recalls the 'wonderful miracle' of the previous spring when his own hands had joined Gertrude's 'in a shockwave passed out through all the galaxy' (*NS* 420). He feels instantly 'mocked and rejected' (*NS* 420). Upset by this turn of events that he misconstrues at the house, he loses his way and allows himself to get completely lost in the mountains at night:

> They were piled round about him, great leaning monumental shafts, wherein the moonlight suggested little steps and ledges and even, close at hand, their spotty crinkly texture. The rocks rose up in tense quietness, like a symphony of frozen inaudible sound. (*NS* 422)

Rocky crags bar his way and a sense of entrapment causes him to panic until he finds himself once again confronted by the Great Face. 'For a moment Tim forgot everything except the marvel that was before him', writes Murdoch (*NS* 424). Next morning, Tim awakens among the rocks to a grey stillness that sheds austere light and effects a 'terrible immobility' on the immediate landscape except on the 'crystal circle of water' which was pulsating with a 'different more urgent rhythm' (*NS* 424–25). He denies himself a drink from the 'forbidden water' and thinks that 'he would like to die' (*NS* 426–28). Adrift in the solipsism of his own tribulations, he senses heroism in swimming to rescue a flailing dog he suddenly sees being swept rapidly over the edge into a subterranean channel.

Only later does Tim have the opportunity to reflect on the high drama of his ordeal: 'The murderous waters of the canal had beaten and baptized him back to life. Had he then returned to Gertrude purged and punished?' (*NS* 486). But Murdoch immediately calls time on this self-indulgent inner monologue, pronouncing reprovingly: 'This was romanticism' (*NS* 486). Innately suspicious of such solipsistic narrative, Murdoch continues in pragmatic vein: 'He had come back because he was bruised and bleeding and half drowned' (*NS* 486). The waters are evidently dangerous, but not 'murderous' (*NS* 486). The rocks do not set out to frighten Tim: the ages-old palpable presence of the mountains figures adamantine proportions but lithic indifference. Murdoch is self-mocking inasmuch as she gently mocks Tim. She illustrates just how easy it is in stressful times to turn inward on one's own concerns, and work to avoid this inclination is never complete. Bennett suggests that allowing for the 'agentic capacity' of landscape is a step towards ecological sensibility but it is 'too often bound up with fantasies of a human uniqueness in the eyes of God, of escape from materiality, of mastery of nature; and even where it is not, it remains an aporetic or quixotic endeavor' (2010, ix). 'Inhuman agency undermines our fantasies of sovereign relation to the environment', insists Cohen (2015, 9). In Murdoch's depiction of the mountains in *Nuns and Soldiers* she reveals a reverential sensibility to the numinous material world that exists in and of itself: our place on Earth subject to chance and fleeting.

Drawing on key novels published over a period of about twenty-five years (1956–1980), this chapter has sought to chart the development of Murdoch's animistic sensibility. Her apprehension of the equity of all life connects to an emerging sensibility to, and respect for, a participative world. Murdoch attends to nature's minutest details and accommodates these as she starts to contemplate vaster scales of time and space. A pattern of pivotal solitary moments of epiphany for key characters in nature emerges. From the garden as an aesthetic space in its interchange with nature as her starting point, Murdoch's representations of vegetal and lithic materiality evolve as she moves towards her more primordial depictions of agentive landscape. John Bayley, writing *Iris: A Memoir* as his wife became increasingly diminished by Alzheimer's disease, remembers that she had always possessed some sort of consideration for the 'life of inanimate things':

> I used to tease her about Wordsworth's flower, which the poet was confident must 'enjoy the air it breathes'. 'Never mind about flowers', Iris would say, impatiently and somewhat mysteriously. 'There are other things that

matter much more'. Though good about it at the time, she also felt real sadness for the abandoned bottles [referring to an earlier tale of their habit of laying empty bottles 'to rest' under stones next to the road], and I think of it now when she stoops like an old tramp to pick up scraps of candy paper or cigarette ends from the pavement. She feels at one with them, and will find them a home if she can. (1999, 119)

It is a poignant fact that Murdoch's powerful sense of animism remained with her long after she had forgotten that she had ever written poetry, novels or philosophy.

References

Primary Sources: Novels

Murdoch, Iris. 1994. *The Green Knight* (*GK*) (1993). London: Penguin
———. 1999. *The Sea, The Sea* (*TSTS*) (1978). London: Vintage
———. 2000. *The Flight from the Enchanter* (*FFE*) (1956). London: Vintage
———. 2000. *An Unofficial Rose* (*UR*) (1962). London: Vintage
———. 2000. *The Unicorn* (*TU*) (1963). London: Vintage
———. 2000. *The Nice and the Good* (*NG*) (1968). London: Vintage
———. 2000. *The Philosopher's Pupil* (*PP*) (1983). London: Vintage
———. 2001. *Nuns and Soldiers* (*NS*) (1980). London: Vintage

Primary Sources: Philosophical Writings

Murdoch, Iris. 1997. 'Against Dryness' (AD). In *Iris Murdoch, Existentialists and Mystics: Writings on Philosophy and Literature*, ed. Peter J. Conradi, 287–295. London: Penguin.
———. 1997. 'The Fire and the Sun: Why Plato Banished the Artists' (FS). In *Iris Murdoch, Existentialists and Mystics: Writings on Philosophy and Literature*, ed. Peter J. Conradi, 386–463. London: Penguin.
———. 1997. 'The Idea of Perfection' (IP). In *Iris Murdoch, Existentialists and Mystics: Writings on Philosophy and Literature*, ed. Peter J. Conradi, 299–336. London: Penguin.

———. 1997. 'On "God" and "Good"' (OGG). In *Iris Murdoch, Existentialists and Mystics: Writings on Philosophy and Literature*, ed. Peter J. Conradi, 337–362. London: Penguin.

———. 1997. 'The Sublime and the Beautiful Revisited' (SBR). In *Iris Murdoch, Existentialists and Mystics: Writings on Philosophy and Literature*, ed. Peter J. Conradi, 261–286. London: Penguin.

———. 1997. 'The Sovereignty of Good over other Concepts' (SGC). In *Iris Murdoch, Existentialists and Mystics: Writings on Philosophy and Literature*, ed. Peter J. Conradi, 363–85. London: Penguin.

———. 1997. 'Thinking and Language (TL). In *Iris Murdoch, Existentialists and Mystics: Writings on Philosophy and Literature*, ed. Peter J. Conradi, 33–42. London: Penguin.

———. 2003. *Metaphysics as a Guide to Morals (MGM)*. London: Vintage.

Primary Sources: Poetry and Other Writings

Murdoch, Iris. 2014. Untitled poem ['The trailing stars tell of dooms'], Poetry Notebook 6 (PN6) and *Iris Murdoch Review* 5 (*IMR*5)

———. 2019. Millionaires and Megaliths. *Iris Murdoch Review* 10 (*IMR*10)

Primary Sources: Journals from the Iris Murdoch Collections

Murdoch, Iris, Journal 3 (J3), 4 June 1945–12 May 1947, KUAS202/1/3.

———, Journal 7 (J7), January 1949–1 January 1953, KUAS202/1/7.

———, Journal 14 (J14), 1 January 1981–8 August 1992, KUAS202/1/14.

Secondary Sources: Books, Chapters, Journal Essays, Podcasts, Radio Programmes, Television Programmes, Websites

Abram, David. 1996. *The Spell of the Sensuous: Perception and Language in a More-Than-Human World*. New York: Vintage.

Bayley, John. 1999. *Iris: A Memoir* (1998) London: Abacus

Bennett, Jane. 2010. *Vibrant Matter*. Durham NC: Duke University Press.

Bolton, Lucy. 2022. Iris Murdoch and Feminism. In *The Murdochian Mind*, ed. Silvia Caprioglio Panizza and Mark Hopwood, 438–450. London: Routledge.

Brooke, Rupert. 1992. The Old Vicarage, Grantchester. In *The Collected Poems of Rupert Brooke*. London: Macmillan.

Brophy, Brigid. 1971. In Pursuit of a Fantasy. In *Animals, Men and Morals*, ed. Stanley Godlovitch, Rosalind Godlovitch, and John Harris, 125–145. London: Gollancz.

Byatt, A. S. 1994. *Degrees of Freedom: The Early Novels of Iris Murdoch*. London: Vintage.

Caprioglio Panizza, Silvia. 2022. *The Ethics of Attention: Engaging the Real with Iris Murdoch and Simone Weil*. London: Routledge.

Cohen, Jeffrey Jerome. 2015. *Stone: An Ecology of the Inhuman*. Minneapolis: University of Minnesota Press.

Conradi, Peter J. 2001. *Iris Murdoch: A Life (IMAL)*. London: HarperCollins.

Crenshaw, Kimberlé. 1995. Mapping the Margins: Intersectionality, Identity Politics, and Violence Against Women of Color. In *Critical Race Theory: The Key Writings that Formed the Movement*, ed. Kimberlé Crenshaw, Neil Gotanda, Gary Peller, and Kendall Thomas. New York: The New Press.

Curry, Patrick. 2014. The third road: Faërie in hypermodernity. In *The Handbook of Contemporary Animism*, ed. Graham Harvey. Abingdon: Routledge.

Dipple, Elizabeth. 1982. *Iris Murdoch: Work for the Spirit*. London: Methuen & Co.

Dooley, Gillian, ed. 2003. *From a Tiny Corner in the House of Fiction: Conversations with Iris Murdoch (TCHF)*. Columbia SC: University of South Carolina.

———. 2022. *Listening to Iris Murdoch*. Iris Murdoch Today. London Palgrave Macmillan.

Duncker, Patricia. 2000. Introduction. In *Iris Murdoch, The Flight from the Enchanter*. London: Vintage.

Haraway, Donna. 1992. The Promises of Monsters: A Regenerative Politics for Inappropriate/d Others. In *Cultural Studies*, ed. Lawrence Grossberg, Cary Nelson, and Paula A. Treichler, 295–337. New York: Routledge.

Harvey, Graham. 2017. *Animism: Respecting the Living World*. 2nd ed. London: Hurst & Co.

Heidegger, Martin. 1971. *Poetry, Language, Thought*, trans. Albert Hofstadter. New York: Harper & Row.

Horner, Avril. 2010. 'Refinements of Evil': Iris Murdoch and the Gothic. In *Iris Murdoch and Morality*, eds. Anne Rowe and Avril Horner, 70–84. Basingstoke: Palgrave Macmillan.

Horner, Avril, and Anne Rowe, eds. 2015. *Living on Paper: Letters from Iris Murdoch 1934–1995* (*LoP*). London: Chatto & Windus.

Iovino, Serenella, and Serpil Oppermann, eds. 2014. *Material Ecocriticism*. Bloomington: Indiana University Press.

James, Henry. 1987. *The Critical Muse: Selected Literary Criticism*. London: Penguin Books.

Jordan, Julia. 2012. 'A Thingy World': Iris Murdoch's Stuff. *Modern Language Review* 107 (2): 364–378. https://www.jstor.org/stable/10.5699/modelangrevi.107.2.0364 [accessed 5 November 2018].

Kermode, Frank, *'Iris Murdoch'*, *Modern Novelists*, BBC 1, 24 May 1965. https://www.bbc.co.uk/archive/modern-novelists%2D%2Diris-murdoch/zknqrj6. Accessed 10 June 2022.

Kuehl, Linda. 1969. Iris Murdoch: The Novelist as Magician / The Magician as Artist. *Modern Fiction Studies* 15 (3): 347–360. http://www.jstor.org/stable/26278908. Accessed 24 October 2022.

Leeson, Miles, Lucy Oulton, and Frances White. 2022. An Unofficial Rose. *Iris Murdoch Podcast*, 6 December 2022.

London Environmental Network. 2022. Environmentalism in Action: Understanding Environmental Justice and Intersectionality. https://www.londonenvironment.net/environmentalism_in_action_environmental_justice_and_intersectionality. Accessed 12 August 2022.

Marvell, Andrew. 1996. The Garden. In *The Complete Poems*. London: Penguin.

Miles, Robert. 2002. *Gothic Writing 1750–1820*. Manchester: Manchester University Press.

Milligan, Tony. 2022. Loving Attention to Animals. In *The Murdochian Mind*, eds. Silvia Caprioglio Panizza and Mark Hopwood, 468–478. London: Routledge.

Moden, Rebecca. 2023. *Iris Murdoch and Harry Weinberger: Imaginations and Images*. Iris Murdoch Today. London: Palgrave Macmillan.

Nicol, Bran. 2004. *Iris Murdoch: The Retrospective Fiction*. 2nd ed. Basingstoke: Palgrave Macmillan.

Oulton, Lucy. 2020. Loving by Instinct: Environmental Ethics in Iris Murdoch's *The Sovereignty of Good* and *Nuns and Soldiers*. Études britanniques contemporaines 59: *Iris Murdoch and the Ethical Imagination–Legacies and Innovations*. https://doi.org/10.4000/ebc.10237

Plumwood, Val. 2014. 'Nature in the active voice' (2009). In Graham Harvey, ed. *The Handbook of Contemporary Animism*. 441–453. Abingdon, Routledge.
Racine, Jean. 1992. *Phèdre*. London: Penguin.
Ripley, Becky, and Emily Knight. 2023. Rivers and the Rights of Nature. *Naturebang.*, BBC Radio 4, 10 January https://www.bbc.co.uk/sounds/play/m001gwz2. Accessed 10 January 2023.
Rose, W. K. 2003. Iris Murdoch, Informally. In *From a Tiny Corner in the House of Fiction: Conversations with Iris Murdoch* (*TCHF*), ed. Gillian Dooley, 16–29. Columbia, SC: University of South Carolina.
Rowe, Anne. 2002. *The Visual Arts and the Novels of Iris Murdoch*. Lampeter: Edwin Mellen Press.
———. 2019. *Iris Murdoch*, Writers and Their Work. Liverpool: Liverpool University Press.
Sale, Roger. 1963. The Miracle Worker. *The Massachusetts Review* 4 (2): 423–429.
Salwa, Mateusz. 2014. The Garden—Between Art and Ecology. *Proceedings of the European Society for Aesthetics* 6: 316–327.
Shakespeare, William. 2008. *As You Like It*, The RSC Shakespeare. London: Macmillan.
Smith, Mick. 2011. *Against Ecological Sovereignty: Ethics, Biopolitics, and Saving the Natural World*. Minneapolis: University of Minnesota Press.
Spender, Natasha. 2019. Nuns and Soldiers: Iris in Provence. In *Iris Murdoch: A Centenary Celebration*, ed. Miles Leeson, 65–72. Devizes: Sabrestorm.
Thompson, Evan. 1995. *Colour Vision: A Study in Cognitive Science and the Philosophy of Perception*. London: Routledge.
Weil, Simone. 1952. *Gravity and Grace*. Trans. by Emma Craufurd. London: Routledge
White, Lynn. 1967. The Historical Roots of our Ecologic Crisis. *Science New Series* 155 (3767): 1203–1207. http://www.jstor.org/stable/1720120. Accessed 26 September 2019.
White, Frances. 2020. Anti-Nausea: Iris Murdoch and the Natural Goodness of the Natural World. *Études britanniques contemporaines* 59: *Iris Murdoch and the Ethical Imagination–Legacies and Innovations* https://doi.org/10.4000/ebc.10212
Wittgenstein, Ludwig. 1958. *Philosophical Investigations*, 2nd ed, trans. G.E.M. Anscombe. Oxford: Basil Blackwell.

4

The Sovereignty of the Sea in *The Sea, The Sea*

'Un Monologue de "Moi"'

This next chapter moves to the coast and to a new perspective on Murdoch's Booker Prize-winning *The Sea, The Sea*, examining what the littoral in this novel literally physically represents. I have said that the reading and writing of poetry inspired and animated Murdoch's novelistic art: poetry-writing provided the driving force behind her significant linguistic accomplishments. It is a 'great joy', she once said, that poetry 'in one language can inspire poetry in another language' (*TCHF* 149). Murdoch took particular delight in the French poets and some of her French verse anthologies now reside in the Iris Murdoch Collections.[1] Murdoch is thought to have particularly admired the symbolist, Paul Valéry. In *The Unicorn*, Marian declares: '*Le vent se lève! … il faut tenter de vivre!*' (*TU* 54) ['The wind is rising! … We must try to live!'] (McGrath

---

[1] St John Lucas (ed.) (1907) *The Oxford Book of XVIII–XXth Century French Verse*. Oxford, Clarendon Press, IML 838; Laurence Adolphus Bisson (1942) *Des Vers du France*. Harmondsworth, Penguin, MLL 59; Alan Boase (1952) *The Poetry of France: from André Chénier to Pierre Emmanuel*. London, Methuen, IML 837, from the Iris Murdoch Collections.

and Comenetz 2013, 13).[2] The young woman produces lines from Valéry's '*Le cimetière marin*' (1922) to express her concern for Hannah's mystifying and worrisome incarceration. Marian's call to action ironically heralds that moment, too, when she becomes aware of a strange paralysis affecting her own desire to step outside the confines of Gaze Castle.

Fifteen years later, the Valéry poem has acquired far greater significance in Murdoch's oeuvre by becoming one of the major sources of inspiration not only for the title of her Booker Prize-winning novel, *The Sea, The Sea* (1978), but also for the ocean present at the novel's centre. This source is confirmed when her narrator, describing the church in the nearby village of Narrowdean, notes the 'most attractive *cimetière marin*, which evidences a more spacious past than one would expect this "one horse" village to possess' (*TSTS* 13). Inspiration for her title is also thought to derive from the joyful cry, 'θάλαττα! θάλαττα!' [the sea! the sea!], uttered from the lips of Cyrus the Younger's sea-loving Greek mercenaries in Xenophon's *Anabasis* on encountering the Black Sea at last after their long march (Dipple 1982, 299). The opening stanza of Valéry's '*Le cimetière marin*' foregrounds the seascape's objective power, assigning equity or impartiality to the noon day sun as the sea carries on its perpetual cycle: '*Midi le juste y compose de feux / La mer, la mer, toujours recommencée!*' ['Noonday the just composes out of fires / The sea, the sea, forever recommencing!] (Valéry 1971, 13). Murdoch adopts this repetition for the title of her novel to give prominence to the material presence and perpetual cycle of the tidal waters at Shruff End. In doing so, Murdoch juxtaposes the timeless and predictable composition and constancy of the northern seascape with a narrator who will prove himself to be profoundly unreliable. Valéry's speaker acknowledges the restive power of the seascape's serene panorama, whose natural force bestows a peace of mind, '*O récompense après une pensée / Qu'un long regard sur le calme des dieux!*' ['After a thought, O then what recompense / A long gaze at the gods' serenity!'] (1971, 13). Valéry's setting evokes the sort of benign seaside idyll that *The Sea, The Sea*'s narrator seems to anticipate, one senses, even purposefully feels entitled to, having made the decision to retire to live on the northern seaboard.

---

[2] McGrath and Comenetz's *Valéry's Graveyard: Le Cimetière marin translated, described, and peopled* is one of two translations I rely on in this chapter.

4 The Sovereignty of the Sea in *The Sea, The Sea*

Valéry once described his poem as '*un monologue de "moi"*' (Dipple 1982, 300), a self-regarding sentiment echoed in the early pages of *The Sea, The Sea* by Murdoch's narrator. Valéry's purpose in composing '*Le cimetière marin*' was simply to 'demonstrate a particular stanza form', according to Bayley: the poet's wider intention being to produce work that 'discredits *vers libre*' (Bayley 1957, 50).[3] Whatever Valéry's project, the poem presents both highly controlled form and a rational clarity in prioritizing the eternal seascape:

Quel pur travail de fins éclairs consume
Maint diamant d'imperceptible écume,
Et quelle paix semble se concevoir!
['How pure, how fine a flashing work consumes
Diamond on diamond of sightless foam,
And what a peace appears to be conceived!']
(McGrath and Comenetz 2013, 5)

While *The Sea, The Sea*'s maritime space appears to conform to the Valérian setting, the novel's first-person narrator's self-conscious, romanticized, and chaotic attempts at literary pastiche in the novel work antithetically to the poem's clarity of form. Murdoch's narrator-protagonist conceives at first a 'memoir', attempts 'word-pictures', then decides he is writing a 'diary' which becomes a 'philosophical journal', a 'formal account of myself', an 'autobiography', a 'novel', a 'novel-diary' and, the least trustworthy descriptor of them all, a 'more deeply reflected and more systematically remembered [...] novelistic memoir' (*TSTS* 1–3, 17, 26, 99, 169, 239). In the end, of course, even *The Sea, The Sea*'s narrator has become untrusting and disdainful of his own work, denouncing its superficiality once and for all as a 'chattering diary' (*TSTS* 483). But before Murdoch reveals her protagonist's writerly chaos, or even his identity, she lures her reader to the sea and its presence, its significant and abundant life, its inherent constancy and, importantly, its transparent indifference to human affairs. The sea serves to emphasize the troubling instability of her narrator's dubious tale from the novel's inception.

---

[3] For further thoughts on the genesis of the poem, see 'Commentary on "The Graveyard by the Sea"' (Valéry 1971, 23–73).

Murdoch conceives a realist coastal environment in *The Sea, The Sea*. The novel's littoral scenes function as (albeit fictional) literal and material spaces. This chapter seeks to broaden the hitherto often purely psychological interpretation made of these waters in order to demonstrate the vital role the 'blue ecology' (Mentz 2009, 1000) abutting Shruff End plays in the events and resolution of the story. Mentz argues that '[l]ooking closely at the sea, rather than just the land, challenges established habits of thought' (2009, 997). With her characters embedded in this blue literary ecology, *The Sea, The Sea* testifies to the integrity and transformative power of the maritime environment. The tidal waters provide a vital site of moral revelation and serve to convey Murdoch's view of the transformative power and mutual benefit of attending to the natural world. The realistic rendering of the novel's material seascape functions to resist the subjectivist belief that outside of human consciousness resides a contingent and chaotic world producing symbolic happenings subject to being continuously ordered by the narrator-protagonist's consciousness. In fact, the seascape is the implacable constant which exposes our human tendency to tell stories, even create fantasies, in an attempt to make sense of and shape the world. The overwhelming presence of the non-human world, in particular the sea, and the taxonomic diversity of both its authorially described and implied constituents operate as a sublimely indifferent force in the novel. The sea counters the claims made upon it by a narrator who is in tumultuous chaos and relies on the natural world's inherent neutrality and constancy to attain order.

At the beginning of the novel, then, Murdoch prioritizes the sea over the identity of her first-person narrator, even as she interposes the 'me as I write' of the narrator in a poetic description of her primary subject at the turn of the tide on a fresh, calm, late-spring day: 'The sea which lies before me as I write glows rather than sparkles in the bland May sunshine', declares the transient memoirist (*TSTS* 1). Her narrator, redolent of Coleridge's 'dread watch-tower of man's absolute self' (1997, 338, l. 40), writes in a state of self-perceived solitary power atop his seaside promontory, casting an eye over the sublime view, and declares that the combined (yet contradictory) purpose of his intended memoir-writing is to 'repent of egoism' and 'to *think* about myself at last' (*TSTS* 3). The narrator's inclusion of self, writing from his imposing 'sea house' as

4 The Sovereignty of the Sea in *The Sea, The Sea*

master of all he surveys, produces an unsettling simulacrum of Caspar David Friedrich's wanderer: with back turned to his observer, the wanderer surveys the foggy landscape from high atop a mountain. The *Rückenfigur* compositional device, more commonly found in visual art and so-called after Friedrich's *Der Wanderer über dem Nebelmeer*, is regarded as a key signifier of Romanticism.[4] The device expresses nature's sublime potential as idealized in the imagination of its human beholder and, serving as a form of proxy, the imperious viewer is lured into sharing his idealized vision of the sublime landscape.[5] Murdoch invokes her *Rückenfigur* in *The Sea, The Sea*'s introductory paragraph, seducing the reader through the sublime quality of the seascape's palette of rich hues and its concomitant enigmatic tranquility, as expressed by her grandiloquent observer/narrator.

Conjuring the Romantics in this way exposes, by another means, Murdoch's ongoing quarrel with the movement. Benjamin Lipscomb believes that Murdoch's disapprobation emerges from Romanticism's tendency to superficiality, specifically of the sort of unexamined nineteenth-century Romantic trope where it may seem 'exhilarating and exalting to stare into the abyss' but 'the seeming gloom of the existentialist conceals elation' (RIP 2018). Murdoch, says Lipscomb, 'gives us permission to laugh at the pretension', and *The Sea, The Sea* in some senses provides the reader with just such an opportunity (RIP 2018).[6] Murdoch has no

---

[4] Caspar David Friedrich, *Der Wanderer über dem Nebelmeer* [*The Wanderer above the Sea of Fog*] (1817) oil on canvas, 980 ×740 mm, Kunsthalle Hamburg. The literal meaning of '*Rückenfigur*' is 'back-figure' but as an artistic device, 'a figure seen from behind'; I argue that Murdoch's protagonist assumes the commanding demeanour of Friedrich's wanderer. See also, Elizabeth Prettejohn, *Beauty & Art 1750–2000* (2005, 54–56).

[5] The Friedrich portrayal of the 'archetypical image of the mountain-climbing visionary' has influenced Western conceptions of mountain climbing ever since the Romantic era, powerfully representing a new idea that reaching a mountain's summit should be something to be admired, according to Robert Macfarlane (2003, 157).

[6] In Lipscomb's book about the four philosophers published three years after his Royal Institute of Philosophy lecture, Lipscomb writes of Murdoch's 'existentialist hero' who, akin to Friedrich's wanderer, 'stares down the abyss, finding sublimity both in the unfathomable drop and in the spectacle of himself standing above it. Because the absurdity of life, not the obscurity of the land below, is the source of the sublime, the existentialist hero isn't a mountain climber but one of the alienated rebels who have been part of the Western canon since the Romantic era', and Lipscomb offers Charlotte Brontë's Mr Rochester as one such example. Lipscomb says, 'Murdoch notes that the

difficulty with Romanticism's canonisation of the natural world; where she does take issue is with the movement's general tendency to centre, impose, even champion the solitary figure in the environment with whom we are expected to be complicit. In *The Sea, The Sea*, even if there is no overt invitation to laugh at her narrator's pretension, his note-to-self to 'pioneer some easier way to get to the tower' (*TSTS* 91), for example, ought to raise a wry smile for our explorer/hero. His strained relationship with his surroundings is, at the very least, baffling. Importantly, Murdoch wants us to dig deep beneath the evident superficiality to see what such affectation conceals.

Murdoch's narrator-protagonist paints, then, an idyllic picture of the maritime scene. The richly romantic palette of colours—'luxurious purple', 'emerald green' with a personified 'cloudless sky […] pale at the indigo horizon which it lightly pencils in with silver'—seduces the unsuspecting reader, when suddenly the sublime acquires a Gothic chill, heralding an unexpected change in mood, at the end of the opening passage: 'But the sky looks cold, even the sun looks cold' (*TSTS* 1). What follows is an oblique reference to a mysterious happening 'so extraordinary and so horrible that I cannot bring myself to describe it', expressed in a short rider to the opening paragraph of what, at this stage, forms the opening of the narrator's proposed memoirs (*TSTS* 1). Murdoch cuts his romantic conception of coastal retreat dead and, with primordial foreboding, foreshadows a pelagic presence out at sea that her protagonist will believe himself to have sighted but, for the moment, feels quite unable to record. Before the narrator introduces himself to his reader, the text has already acquired a certain unstable quality. What is firmly secure from the start of the novel, by invocation of the Valéry poem, is the constancy of the natural seascape. This constancy juxtaposes the unreliability of *The Sea, The Sea*'s storyteller whose solipsistic nature has been made abundantly clear through the allusion to Friedrich's paradigmatic landscape.

---

invitation to identify with this type involves more than a little self-aggrandizement: […] the gloom characteristic of existentialist writing "is superficial and conceals elation" (NM 50)' (Lipscomb 2022, 128–29).

4 The Sovereignty of the Sea in *The Sea, The Sea*

Murdoch's decision to prioritize the sea lures the unwitting reader into a sympathy for the attempt at memoir that follows. However, John Mullan reminds us that '[c]onfession has long been a form in which fiction is cast. Such narrative engages us not simply by giving access to a character's thoughts' and in 'opening a gap between the "I" who tells the story and the "I" who is the past self' lies the 'drama of a person trying to make sense of him- or herself' (Mullan 2006, 45). Mullan suggests we ought to be particularly circumspect about those narrators who appear to possess total recall of another's words. In doing so, he reminds us of the use writers 'from *Moll Flanders* to *Lolita*' have made of first-person narration, as 'the means of drawing a reader into some disturbing sympathy with a character's misdeeds' (Mullan 2006, 45). For example, in the first of these novels, it is the settings of England and Virginia that are faithfully drawn because, if Defoe (through Moll Flanders) 'wants to be believed to be telling the truth when he isn't', he has to 'mix his untruth with a good deal that really is true' (Jones 1993, vii). *The Sea, The Sea* is constructed to function in similar vein when the narrator claims his account 'is curtailed, but omits nothing of substance and faithfully narrates the actual words spoken' (*TSTS* 239). As we, as readers, become unsettled and begin to mistrust the novel's narratorial perspective, integrity must be found elsewhere. In this novel, it is the seascape that provides truthful admixture to this otherwise unstable storytelling and, of course, secures Murdoch's place in the long tradition of the sort of narrative circumspection that Mullan describes.

Ecocritical theory offers a contemporary explication of what the constancy and integrity of Murdoch's sea might mean for the novel, primarily through a consideration of the novel's narrative form. Recently, ecocritics have become alert to the relevance and rich complexity of narrative structure, which succeeds the previously longstanding commitment to prioritizing a work's content over its form. Erin James and Eric Morel argue that, until recently, a 'consideration of narrative structures has tended to stay separate from and in a subservient role relative to ethical discourse in much ecocritical scholarship' (2018, 357). Meanwhile, Ursula Heise's interest in the 'aesthetic transformation of the real' is founded on the promise of its potential to shift the 'ecosocial imaginary' toward a greater environmental awareness in the reader (2010, 258). Similarly, bioculturalist Nancy Easterlin has expressed her interest in the 'agentive force' of narrative, whereby

> integrating the actions and purposes of human groups with their prescribed domain, [...] brings into relation and coordinates sequence, causality, physical place, knowledge of interaction with human others, and self-concept. (2012: 139)

Easterlin considers the potential for a more sophisticated interpretation of a given text's narrative structure, which can then lend itself to greater ecological consciousness. Erin James designates 'econarratology' to this pairing of narrative art with environment, her particular interest lying in 'what happens to readers when they mentally inhabit the imagined environments' expressed in such narratives (2018, 359). James describes the sort of 'storyworlds' that

> readers simulate and transport themselves to when reading narratives, the correlations between such textual, imaginative worlds and the physical, extratextual world, and the potential of the reading process to foster awareness and understanding for different environmental imaginations and experiences. (2015, xv)

In terms of *The Sea, The Sea*, and of particular interest to Murdoch scholarship, must be the strategies and devices created by Murdoch to encourage her reader towards a form of attachment to the imaginary seascape, whose role in this story is that of a truth-telling omni-sea. In other novels, Murdoch asserts geographical specificity on the seascapes, for example, the Dorset coast in *The Nice and the Good*, Cumbria in *Nuns and Soldiers*, and Sandycove, Dun Laoghaire, Ireland, in *The Red and the Green*; whereas the location of her 'blessed northern sea' (*TSTS* 2) in *The Sea, The Sea* remains indeterminate, thereby creating this sense of an omni-sea or every-sea. Some environmentalist discourse would see fit to discount such indeterminacy were it not for Robert Kern's ground-breaking essay in which he argues that all texts are at least potentially environmental [...] in the sense that all texts are literally or imaginatively situated in a place, and in the sense that their authors, consciously or not, inscribe within them a certain relation to their place. (2000, 10)

Heise argues that the 'basis for genuine ecological understanding [...] lies in the local' (2008, 28), although this would not appear to account for fictional locations in quite the same way. Furthermore, Murdoch renders her ocean significant not for its geographical place but for its fully realized and inhabited ethical space. As readers of the novel, we have been accustomed to discussing the degrees of credibility of her protagonist, but what has not been fully considered before is how his unreliability is productively and antinomically mirrored by the constancy of the novel's richly inhabited seascape.

'I Am My Well-Known Self'

Having introduced the ocean that is proving to be an unwelcome distraction, Charles Arrowby introduces his autobiographical self: 'I am my well-known self, made glittering and brittle by fame' (*TSTS* 3). In the physical world, he is pictured strikingly alone: 'wifeless, childless, brotherless, and sisterless' (*TSTS* 3). Charles has relinquished the inauthentic or, at best, make-believe world of the theatre for a reunion with his spiritual self, achievable, so he believes, through a contemplation of the natural world and in the writing of his memoirs. Here, he plans to record his '*pensées* against a background of simple descriptions of the weather and other natural phenomena' (*TSTS* 2). He wants to render a certain permanence to his writing, having deliberately resisted such 'personal and reflective' exposure to the craft during his time as a theatre director. 'What I wrote before was written in water and deliberately so', and he claims to want to record his reflections as he 'learn[s] to be good' (*TSTS* 2).[7] Again, Murdoch is gently mocking the sort of Romantic conceit that Bayley

---

[7] Charles alludes to the inscription on John Keats's gravestone at the *Cimitero Acattolico* in Rome: 'Here lies one who was writ in water', at once signalling the romanticization of his own circumstances—the transient nature of fame—and, more generally, the evanescence of human life.

describes as being aware 'of the isolated creating self on the one hand, and of a world unrelated, and possibly uninterested and hostile, on the other' (1957, 10). Charles is the emotional 'creating self' demanding an authentic source of aesthetic experience from a sea that proves to be 'unrelated, and possibly uninterested' and certainly is, in his view, 'hostile' (Bayley 1957, 10).[8] It is the 'undoing of this false ideal' of the littoral idyll that is a key purpose of *The Sea, The Sea* for Dipple (1982, 299).

In Dipple's view, the Murdochian seascape carries rich symbolic power and psychoanalytic purpose. She airily codifies the maritime scene with a sort of barren blankness in support of a symbolic interpretation for psychoanalytic purposes, whereby a 'detached, neutral and empty' sea functions as a 'vast empty force which resists all imprints' (Dipple 1982, 299). Charles's mind, she claims, pours 'questionable content' into Murdoch's 'empty sea of reality' (1982, 301). For Dipple, then, the sea adds both metaphorical and psychological significance to Charles's tale, its presence in the novel functioning as the representation of his turbulent mind and chaotic relationships. She is, of course, not alone in foregrounding the sea's symbolic value. Rowe regards the sea in *The Sea The Sea* as Murdoch's 'most sustained attempt at extending the boundaries of language to accurately portray the amorphous nature of human consciousness, for which the sea itself is the central symbol' (2019, 28). As I have said, Murdoch engages with the natural world in similar vein to her engagement with visual art when words prove inadequate, and both Dipple and Rowe figure the natural turbulence of coastal waters as mirroring the inchoate qualities of human consciousness, a symbolism that serves to 'make these inner features visible' to the reader, Rowe explains (2019, 50). Bove anthropomorphizes the sea as 'fierce and alien' and contends that the

---

[8] In her book on Murdoch's visual acuity and intellectual relationship with painter Harry Weinberger, Moden describes Charles Arrowby as 'intuitively aware that he should turn his attention outwards to try to see the infinitely coloured beauty of nature, but his painstakingly considered vocabulary choices (the sea "glows rather than sparkles", it is "opaque however, not transparent") and his carefully constructed sentences, balanced for rhetorical effect ("the sky looks cold, even the sun looks cold") produce a rather ostentatious, cerebral description. Charles makes a telling reference to the "surface skin of colour" (*TSTS* 1) on the sea', she says, 'which implies that his impoverished consciousness has interposed a barrier between himself and his surroundings', (Moden 2023, 175–76).

4 The Sovereignty of the Sea in *The Sea, The Sea*

'only creatures Charles had seen were the monsters in his imagination' (1993, 92) when suddenly towards the end of his tale he notices four seals in the water beneath him.[9] It is true that, early on in his description of Shruff End's promontory, Charles records his view of a 'remarkably empty tract of sea', but the emptiness he discerns relates very clearly to a lack of human activity, to the dearth of sea-going vessels in the bay now that fishing boats have abandoned the 'very pretty little harbour'; the 'magnificently built crooked stone quay' is now disused and 'all silted up' (*TSTS* 12). Again, Charles remarks on the ocean's vastness: 'how empty, this great space for which I have been longing all my life' (*TSTS* 15). The perceived emptiness of the sea in the novel represents what Margaret Cohen refers to as a 'sublimation of the sea', a process that occurs both in literature and in the visual arts of the Romantic period (Cohen 2010, 106). Cohen illustrates her argument with another Friedrich painting, *Der Mönch am Meer*—all one can see is sea and sky, and a figure close to the water's edge; this time, relative to the small, partially obscured, solitary *Rückenfigur*, the landscape is vastly more expansive.

Der Mönch am Meer is presented 'as empty elements to be filled with the viewer's imagination' (Cohen 2010, 119), although the perspective that commands the work is inevitably that belonging to the monk figure.[10] Citing Jonathan Raban, Cohen says that Romanticism's 'invention is rather a shift from a depiction of the ocean in the intimacy of human practice to the ocean as "space itself"' (Cohen 2010, 117, cites Raban 1993, 8). She explains that 'novels with oceangoing themes' have tended simply to be treated 'as allegories of processes back on land' (Cohen 2010, 14). Cohen's assertion correlates with the inclination of some critics to categorize the sea in *The Sea, The Sea* as the allegory of the processes in Charles's mind. The disregard for maritime activity, sealife or, indeed, the

---

[9] Charles does not see sea lions in the water (Bove 1993, 92); sea-lions are not native to the British coastal waters that provide the setting for Murdoch's novel. Both marine mammals are classified as 'Pinnipeds' but their apparent similarity arises from evolutionary convergence: seals belong to the *Phocidae* family and are thought to descend from a land-based weasel; sea lions are *Otariidae*, having bear-like ancestors.

[10] Caspar David Friedrich, *Der Mönch am Meer* [*Monk by the Sea*] (1809), oil on canvas, 110 mmm × 1715 mm, Nationalgalerie, Staatliche Museen zu Berlin.

sea itself is what Cohen calls 'hydrophasia' after Allan Sekula, who characterizes the then prevailing twentieth-century oversight as 'forgetting the sea' (Cohen 2010, 14, cites Sekula 2002, 48). Charles's persistent attempts at an imitative evocation of the empty seas of the Romantic sublime seems, at times, to have beguiled critics too.

'How Can Such Gentle Defenceless Animals Survive?'

A close reading of the descriptions of Shruff End's 'strangely unfrequented' tidal waters, constituting a 'real sea with clean merciful tides', reveals in fact a plenitude of non-human creatures that belies the barren waters that critics suggest, and on which symbolic interpretations appear to rely (*TSTS* 440, 2). The 'sea is noisier today and the seagulls are crying' according to Charles the next morning and, during the course of the novel, he vividly chronicles a diversity of marine and coastal creatures, despite his original intention to craft a memoir of his days as a London theatre director (*TSTS* 15). Immodestly declaring himself to be 'no White of Selborne', Charles is captivated by the warm waters in the rock pools, examining them for living organisms, and later noting their biodiverse 'brilliant lucid little civilisations' (*TSTS* 2, 140). He observes lugworms, records sightings of 'cormorant', 'yellow-beaked gulls' and, so he surmises, 'gannets' (*TSTS* 3, 78, 243). At the start of the novel, as he imagines he sees a sea-monster out in the sun-drenched waters, Charles had just that moment been intent on 'watching a remarkably long reddish faintly bristly sea-worm which had wreathed itself into curious coils prior to disappearing into a hole' (*TSTS* 19). Cormorants turn out to be shags, their identification confirmed by his cousin and birding enthusiast, James, who also notices a 'huge rock out in the sea covered with guillemots', and later spots 'choughs, and oyster-catchers', and hears 'a curlew round in the bay' (*TSTS* 321, 387). To the ecological mind, the representation of aquatic birds in the seascape infers the presence of fish, cephalopods and other marine invertebrates.

4 The Sovereignty of the Sea in *The Sea, The Sea*

Charles records his fascination with the amphibians and arthropods that inhabit the seascape, such as a 'whole flotilla of blue flies' that 'crawl upon the surface tension' of the water, not to mention a 'most engaging toad' that has wandered into the larder and that Charles returns to its presumed habitat 'across the wood to the mossy boggy pools beyond the rocks' (*TSTS* 11, 24). 'How can such gentle defenceless animals survive?', he asks with notable sensibility (*TSTS* 24). This occasions him to linger to observe 'the red-tufted mosses [...], mare's tails [...] and that weird yellow flower that catches flies' (*TSTS* 24). Charles senses that '[t]here ought to be seals' as he 'scan[s] the water in vain', instead a 'shadow-cormorant skims the glycerine sea' and the 'rocks are thronged with butterflies' (*TSTS* 67). Charles becomes habituated to his observations; he does grow in attention to the more-than-human, and yet he betrays his proclivity for sovereign pride in his new compulsion to 'go out and study my rock pools' (*TSTS* 88). He believes he is 'becoming more observant' and here he records 'a colony of delightful very small crabs [...] and some ferocious-looking tiny fish with whiskers which resemble miniature coelacanths' (*TSTS* 88). Later still, from 'my cliff' he can see 'tall dark trees of seaweed gently waving and fishes swimming between them' (*TSTS* 139). In place of a depiction of an empty sea, intrinsically valueless and dead, Murdoch creates a convincingly rich and diverse coastal biota in her narrative. The evocation of Gilbert White, here, induces readers to notice and afford significance to the more-than-human lives depicted in the novel. Charles becomes entranced by his surroundings but remains intent on asserting his self-perceived human primacy, prefiguring not only his interactions with his maritime environment but also his wider human relationships.

Conradi's vision of *The Sea, The Sea*'s 'wondrous seascapes' is one that reaches beyond symbolism (*S&A* 297). While he accepts that the littoral setting signifies one of the novel's 'concrete metaphors', he nonetheless asserts a 'literal' force in the sea's presence (*S&A* 312). He powerfully reminds us that the 'world which lies beyond the realm of images, from which all form comes and to which it returns, is partly figured by the sea' (*S&A* 315). He contends that,

> [g]iven the variety and fertility of the sea's many states, moods, and colours, it seems tactless to colonise it as a 'literary device'. Its point, in fact, is its resistance to human devices, its miscellaneous nature, its hugeness and unpredictability. (*S&A* 316)

Conradi recognizes, then, that the realism of the Murdochian ocean functions with alien indifference to human affairs. Nevertheless, in his appraisal of the novel, Conradi also regrets that much of the sea's 'glorious physical description must be left aside' (*S&A* 315). Yet, to leave aside the sea's literary representation in the novel is to downplay the ethical function of its descriptions for the reader. Just as Murdoch's characters 'take on an uncanny actuality in readers' imaginations' (Rowe 2019, 10), so should the realism of her representations of the more-than-human be afforded sustained scrutiny: in this instance, the intertidal biota and the sea itself.

Johns-Putra asks whether it is our whole approach as critics of literary works that is at fault: 'What if […] our ways of reading texts are based on a flawed understanding of identity, empathy, and ethics?' (2021) She speculates on the potential of narrative description to 'provoke readers into acknowledging their places in the world as basically multi-centered and their voices as inherently dialogic'; the big challenge is, in her view, 'to dislodge humans from their long held position of exceptionalism' (Johns-Putra 2019b). For this reason, a new and productive reading of *The Sea, The Sea* is called for, that places Murdoch's depiction of the richly populated coastal waters at its centre and seeks to resist the sovereign exceptionalism displayed by her narrator-protagonist. The sea in *The Sea, The Sea* demands thoughtful attention and critical consideration for what it really represents: itself.

The Millennial Work of the Sea

The starting point, then, must be the alignment of views between Murdochian philosophy and contemporary ecocriticism. Laurence Coupe explains that the 'point is to learn from nature, to enter into its spirit, and to stop trying to impose upon it the arbitrary constraints

4 The Sovereignty of the Sea in *The Sea, The Sea*

which result from our belief in our own importance' (2000a, 1). Coupe's thinking aligns with Murdoch in her germinal essay of 1970, where she attests to the natural world as an 'accessible site of moral change' (SGC 370). It is here that she articulates the important role of nature in the process of unselfing wherein we are capable of shocking ourselves into an awareness of a contingent reality. Arguing for the ethical potential of natural beauty later on in *Metaphysics as a Guide to Morals*, she declares: 'Aesthetic experience is by definition unpossessive' (*MGM* 347). Given that it is a novel that is under discussion here, the sea is inevitably a 'natural scene mediated by culture' (Mitchell 2002, 5). Murdoch's position works to decenter the notion of human sovereignty over a multispecies environment and her novelistic work hints at the universal benefit of such an ethical position.

Lawrence Buell, one of the pioneers of ecocriticism, has his own set of criteria for texts that attempt to represent the natural world. He argues against literary criticism's anthropocentrism. In examining art's representation of the natural world, there has been an overreliance on, or overworking of, symbolism, reducing the natural world's representation to little more than 'textual function', and thereby 'effac[ing] the world' (Buell 1995, 5). He says that, 'environmentally oriented' texts tend to be explicit in their non-human representation or convey legitimate interests beyond human ones, they depict the natural environment as a process rather than a static backdrop or manifest a sense of human accountability for the environment (Buell 1995, 7). These are the criteria of Buell's catalogue of demands for ecological representations: few works would fail to qualify on one account or another.

In *The Sea, The Sea*, Murdoch conceives of the marine environment as a symbiotic process, as a living ecosystem comprising not only her human characters but fully realized non-human representation functioning independently of this human interest. Far from simply constituting a framing device for the tale, the littoral edge serves as a presence that expresses the 'human history [that is] implicated in natural history' (Buell 1995, 7), and which serves to convey the impactful but transitory nature of humanity in a wider planetary context. The novel, in this sense, comprises a number of temporal layers. Murdoch draws a range of scalar dimensions to her reader's attention: we learn much about Charles's childhood and family background in the 'Prehistory'; the '*cimetière marin*' in 'Narrowdean' (or '*Nerodene*') suggests a 'more spacious past', as Charles remarks, which is conveyed by 'Dummy's' gravestone, that records a short life lost at the turn of the twentieth century (*TSTS* 13); and later when Charles is waiting in the church, he notes how time 'had suffered a profound disturbance, and I could feel all sorts of dark debris from the far past shifting and beginning to move up towards the surface' (*TSTS* 131). Dipple is right to sense, however, a much deeper past in the novel, as if 'one feels one is in an ancient world where constant converse with the divine is close at hand through nature, places, and certain human beings' (1982, 302–3). A perception of the primordial is never far out of reach and, at certain moments, is clearly discernible. When Charles seeks to rationalize the sighting of an apparition at Shruff End, he conjures cousin James's allusion to the pre-Socratic philosopher and the forefather of animism, Thales of Miletus. Charles remembers being told by James that: 'Everything is full of gods' (*TSTS* 69). Thales believed that water was the beginning of everything, the $\alpha\rho\chi\eta$ or arche, because water 'exhibits itself naturally to the senses [...] in the three forms solid, liquid and gaseous, as ice, water and steam' (Guthrie 2012, 24). There being no direct record of his philosophical work, Thales's ideas have been handed down verbally and interpreted by Aristotle, among others. Bayley notes in his memoir of Murdoch that many of her characters 'are accompanied by water, as if it were their native environment: the story of their spirits seems to arise from sea or river and return to them' (1999, 127). Murdoch's affinity with water seems to have stemmed from certain such primordial associations. Rachel Carson adds still greater significance to Thales's thinking on water and to Murdoch's own conception of the primordial quality of the element. She avers that the seashore is the 'primeval meeting place of the elements of earth and water, a place of compromise and conflict and eternal change', the place 'of our dim ancestral beginnings' (2015, 8). As 'geological events are reckoned', Carson explains, 'the present relation of land and sea [was] achieved perhaps no more than ten thousand years ago' (2015, 47–8). The findings of this renowned environmentalist can only add value to a new interpretation of *The Sea, The Sea*.

Johns-Putra emphasizes the immense scalar shift in time that today's readers of novels have to apprehend. In doing so, she recalls Morton's

4 The Sovereignty of the Sea in *The Sea, The Sea*

'hyperobjects' (2013, 1): 'things that are massively distributed in time and space relative to humans' (2019b, 249), a new dimension of thinking in the Anthropocene. For Morton, these refer equally to the presence in the universe of black holes or the presence in the world of long-lasting, human-manufactured plastic, that is, the sum of all plastic. Hyperobjects demand a gear-change in time-and-space perspective. More accessible, Johns-Putra suggests, is Clark's 'analysis of the challenges facing the cultural and creative imagination' in the current age (2019a, 249). Johns-Putra explains that

> the disparity between, on the one hand, the immediate and individualised concerns of the discourse of modern human life, including of conventional novelistic narrative, and, on the other hand, the extreme temporal and spatial dimensions of human impacts creates what [Clark] calls 'derangements of scale'. (Johns-Putra 2019a, 249, cites Clark 2012, 148–66)

Vertiginous temporal layering is embraced in *The Sea, The Sea* and signalled by the careful naming of the novel's three sections: 'Prehistory', 'History', 'Postscript: Life Goes On' (*TSTS* vii). Murdoch embeds her human story in a fully inhabited seascape that includes ancient rock and stones that reflect deep geological time: 'exquisitely smooth forms lightly dinted and creased by the millennial work of the sea' (*TSTS* 242), and stars that conjure still deeper astrophysical time.

A similar conception is present in *Nuns and Soldiers*: Guy Openshaw has died, and Gertrude Openshaw and Anne Cavidge are by the sea in Cumbria, where the 'beach of stones [...] had been clashed and beaten by the millennial sea into a terrible density and an absolute smoothness' (*NS* 100). In the case of *The Sea, The Sea*, no destination is provided by the end of the novel. Much like the perpetual lapping of the waves and turns of the tide, the novel achieves a sort of circularity: 'Life Goes On' (*TSTS* vii). We are left not knowing whether Charles simply returns to his old ways but, in some sense, we find ourselves participating in his 'shuddering sense of utter solitude, [his] vulnerability, among these silent rocks, beside this self-absorbed and alien sea' (*TSTS* 101). But we are also invited to look at ourselves: 'What an egoist I must seem in the preceding pages. But am I so exceptional?', asks Charles (*TSTS* 482). Murdoch's purpose is to juxtapose a tale about the solipsistic, sovereign tendencies of her narrator-protagonist with a temporal dimension, replete with ancient stars and brief glimpses of other galaxies, that emphatically lays bare the fleeting nature of life and our human insignificance.

Importantly, Thales is said to have parted company with the Greek mythology of his forebears in an attempt to explain planetary objects and phenomena with natural theories. By acknowledging Thales in this way, Murdoch offers scope for plausible explanations (even if few are explicitly given) for many of the supernatural events of the novel including the manner in which James saves Charles from the seething whirlpool known as Minn's Cauldron. Later, Charles recovers his hidden note about the happening which stretches the reader's credulity when we learn that James has managed to lift his cousin to safety from the '16–20 foot' deep 'churning pit' and when it dawns on Charles that he had, after all, 'relied upon his [cousin's] presence in the world' (*TSTS* 465, 473). Carson offers a natural theory that might counter a supernatural interpretation of this extraordinary feat when she observes that, 'Waves entering a confined space always concentrate all their tremendous force for a driving, upward leap' (2015, 115). At the end of the novel, Charles seems inclined to explain James's untimely death not as a result of James exceeding his powers as a Buddhist, but rather more simply from heart attack: 'There are no mysteries after all', says Charles somewhat unconvincingly (*TSTS* 487).

Kinship

From the start of the novel, Charles proves himself to be averse to any sort of mystical interpretation of events and resistant to any conception of an animistic universe. He sees any such notion as belonging to the theatre or to the imaginary:

> Perhaps I have been surrounded by little gods or spirits all my life, only the magic of the theatre exorcized or absorbed them? […] Well, I have never gone in for persecution mania and do not propose to start now. (*TSTS* 69)

4 The Sovereignty of the Sea in *The Sea, The Sea*

He sets himself the sovereign task of taming his surroundings: 'until I can get some handholds fixed I think I shall shun my "cliff"' (*TSTS* 15); he wants the rock to submit to his purposes. Charles discerns a right of possession: 'I have been out picking flowers upon my rocks' (*TSTS* 74). Still, he seems mildly impressed by the seascape's natural power: 'It is remarkable how quietly firmly powerful my sportive sea can be!' (*TSTS* 6). Charles is intent on possessing and dominating the littoral environment, just as he attempts to execute control over his relationships with the people around him.

Thus far, I have made the case for enlarging the critical treatment of *The Sea, The Sea* beyond the symbolic interpretation of the seascape. However, to the extent that the expression of 'relationship' can also denote 'the bond between the human and nonhuman estates', an exploration of metaphor in the novel can prove powerfully productive (Buell 1995, 180). Buell explains that

> One of the dramatic developments in postromantic thinking about nature has been the decline and revival of the kinship between nonhuman and human. Its metaphysics withered in the last half of the nineteenth century; high modernism announced its death; modern ecologism has brought it back. (1995, 180)

The notion that there should exist a relationship or interaction with the environment and a form of kinship between species has re-emerged as a pressing area of discussion and investigation in our contemporary ecologically fragile age, finding previous iteration in the work of the Romantics. Blake believed in 'nature as a mode of revelation' (Coupe 2000, I, 13). Buell exemplifies Blake's imagined 'complete interchangeability' (1995, 180) of creature and human in the poem, 'The Fly' from his *Songs of Innocence and Experience*, which expresses a poetic regard to sentience and the concomitant equality of right to life that modern ecologism would endorse:

> Art not I
> A fly like thee?
> Or art not thou
> A man like me? (Blake 2009, 50)

At the same time, Buell remarks on Romanticism's historical tendency to put personification under strain until Ruskin called for the 'unencumbered rendering, of the specific characters of the given object, be it man, beast, or flower' (Buell 1995, 188) in the aspiration toward artistic distinction. Projecting human emotion onto the non-human was at issue. Wordsworth with his 'They call upon the hills and streams to mourn,/ And senseless rocks, nor idly' (2007a, 48, l. 478) also appears somewhat contemptuous of his fellow poets through the wanderer's voice, and serves only to emphasize once again the human tendency to self-obsession that also provokes Murdoch's dismay. Although Wordsworth might wander 'lonely as a cloud' (1965, 178), he would likely still view as nonsense the idea that primordial rock formations, for example, might have the capacity to feel how the poet feels. Ecocriticism adopts the Romantics' notion of kinship to interrogate the interaction of human with the more-than-human.

This contemporary position on kinship between human and the more-than-human, newly expressed in modern ecologism, then, offers some insight into the portrayal of the proprietorial and sovereign narrator-protagonist Charles; committed as he may be to Wordsworthian 'recollection in tranquillity' (*TSTS* 1) at the edgelands of the northern seaboard.[11] He begins to document the detail of his surroundings:

> There are many V-shaped ravines containing small pools or screes of extremely varied and pretty stones. There are also flowers which contrive somehow to root themselves in crannies: pink thrift and mauve mallow, a sort of white spreading sea campion, a blue-green plant with cabbage-like leaves, and a tiny saxifrage thing with leaves and flowers so small as almost to defeat the naked eye. (*TSTS* 5)

But he generalizes 'an ugly coast':

> The rocks, which stretch away in both directions, are not in fact picturesque. They are sandy yellow in colour, covered with crystalline flecks, and are folded into large ungainly incoherent heaps. Below the tide line they are

---

[11] Charles uses Wordsworth here to convey a romanticized sense of his circumstances: 'All good poetry is the spontaneous overflow of powerful feelings: it takes its origin from emotion recollected in tranquility', (2007b, xxxiii).

festooned with growths of glistening blistery dark brown seaweed which has a rather unpleasant smell. (*TSTS* 5)

The narrator's vituperative tone unites both passages: he is intent on overcoming (what he perceives as) material intention. Here, as suggested by the lexicon Murdoch selects for him—'contrive', 'incoherent', 'festooned', 'unpleasant'—Charles assigns a hostile form of agential intentionality to the seascape. He seems to believe that the seascape conspires to taint his profiting from it and his consequent enjoyment (*TSTS* 5). He sneers dismissively at the 'slightly sinister appearance' of the weathered rock formations which 'from the swimmer's-eye-view' he is unable to 'dignify with the name of caves' (*TSTS* 5). Then he positions himself again upon a vantage point, this time upon an 'arched bridge of rock' commanding a view whence he derives 'curious pleasure' as he observes the 'violent forces which the churning waves, advancing or retreating, generate within the confined space of the rocky hole' (*TSTS* 5). Charles takes to swimming in the sea but struggles to return to shore. He finds an iron post and after the 'undeserving rope' floats away, he resorts to tying together old curtains to tame access to the water, to render the sea swimmable (*TSTS* 27). A key feature close to the house is the rock bridge that encloses Minn's Cauldron on the seaward side, and Charles derives satisfaction from watching the waves 'killing themselves in fits of rage' (*TSTS* 28). Unable to subdue his new surroundings and render them serviceable, Charles sets himself up in opposition, as if in relational conflict, to the wild environment. Naturally indifferent to all attempts at subjugation, the sea proves to be Charles's nemesis. The narrator-protagonist's efforts to force the coastal waters into submission and his attempts to exert human sovereignty mirror his coercive and controlling behaviour, in particular, towards the women of the novel.

Derek Mahon has composed an Anglophonic poetic interpretation of *Le cimetière marin* for the twenty-first century titled 'The Seaside Cemetery (after Valéry)' (Mahon 2005, loc. 764–846): a rendition of the Valéry poem that blurs the distinction between translation and creative interpretation. By allowing himself to be less constrained by the original poem, Mahon seems able to navigate closer to Valéry's meaning and, in doing so, facilitates a broader understanding of Valéry's influence on *The*

Sea, The Sea. Mahon's poem echoes Valéry's, 'Amour, peut-être, ou de moi-même haine?' (1971, 18–19). In the soul-searching line, 'Self-love, maybe, or hatred of myself?' (Mahon 2005, loc. 764), Mahon's speaker appears to ruminate upon the same sort of complex, often diametrically opposed, moral compulsions that Murdoch combines into her rendering of Charles, one of her most captivating but confounding characters. 'The 'irrefutable worm' that has long since ravaged the corpses of the graves, now never leaves Mahon's speaker as he troubles away at the 'self-love, self-hatred' of his own consciousness:

> Self-love, self-hatred, what's the difference?
> Its secret mordancy is so intense
> the silent gnawing goes by many names.
> Watching, desiring, nibbling, considering,
> it likes the flesh and, even in my dreams,
> I live on sufferance of this ravenous thing. (Mahon 2005, loc. 828)

Charles embodies Murdoch's moral position that 'unobservable inner experience, not existentialist choice, shapes human behaviour' (Rowe 2019, 50), which is why the seascape seems so readily to lend itself to psychoanalytic interpretation when critics see it as reflecting the hidden dark turbulent depths of the human mind. Rowe identifies in Charles, 'repressed feelings of jealousy' and a 'deeply buried fear of women', concealed by 'his virile language and domineering control over them' (2019, 50). The sea monster that Charles claims to have seen early on but can only bring himself to record later in his 'Prehistory' is viewed, then, as a metaphor 'for the voracious Freudian energy of Charles's misogynistic, yet needy, desire for power over women' (Rowe 2019, 50). Rowe interprets the sea monster as an encrypted illustration of Charles's tendency to 'see women only as projections of his unconscious fears and fantasies' (2019, 51). In the Mahon poem, Valéry's mythological sea serpent has been unleashed and demythologized, and becomes a monster-sea, 'drunk on your blue / skin, chewing for ever your own glistening tail / in a perpetual, silent seeming turmoil' (Mahon 2005, loc. 839). The propensity to violence of this vain, misogynistic, strangely obsessed, former theatre director is revealed in his attempts to harness the sea and subjugate his environment, drawing the reader's attention to his aggressive and domineering behaviour towards the women in the narrative.

'An Urgent Dark Desire to *Act*'

The cousins' conceptions of the marine environment are suggestive of their contrasting characters. For Charles, the sea conjures fear and loneliness in death but, to his cousin James, it is 'refreshment to the spirit' (*TSTS* 56). James is one of Charles's '*sinister* witnesses from the past', which infers a personal history that Charles prefers to conceal (*TSTS* 57). He envies his relatives, specifically Aunt Estelle who meant 'freedom, gaiety and noise' to him, and who 'lived in a land which I was determined to find and to conquer for myself' (*TSTS* 60). Charles confesses to exercising his power as a successful theatre director to cover up his failings as an actor and writer, but also admits to wielding this power inappropriately:

> If absolute power corrupts absolutely then I must be the most corrupt of men. A theatre director is a dictator [...] I fostered my reputation for ruthlessness [...] Actors expected tears and nervous prostration when I was around [...] Of course the girls wept all the time. (*TSTS* 37)

One of his 'girls' and a former lover, Lizzie, appeals to him: 'Can we not love each other and see each other at last in freedom, without awful possessiveness and violence and fear, now we are growing old?', which Charles dismisses as a 'silly inconsistent woman's letter' (*TSTS* 47). He recalls being touched by her 'superb obedience' and claims she 'wanted me simply to use her for my happiness', admitting that the 'remembrance of her sufferings often filled me with a kind of tender joy' (*TSTS* 50, 51, 52). A visit to Shruff End from Lizzie arouses an 'old wicked possessive urge' in him while she, according to Charles, has acquired the look of a 'noble girl facing execution' (*TSTS* 96–97). Charles's propensity to abuse is complicated by the anxiety he feels about the relationship with the cousin he so evidently resents: 'The chagrin, the ferocious ambition which James, I am sure quite unconsciously, prompted in me was something which came about gradually and raged intermittently' (*TSTS* 63). In Charles's mind, James has always been his great rival: 'one of us must inhabit the real world, the other one the world of shadows' and he has always been bitterly jealous of him (*TSTS* 57). Charles declares in rather sinister fashion that, after a few years in the theatre, his '"will to power" was bringing in results' and he likens soldier and Buddhist James's 'road to power and glory' in the army to his own 'opening skirmishes in my long war with Clement' (*TSTS* 64).

At first, Charles pronounces 'the main theme' of his memoir to be his deceased lover, Clement, even though he adds bitterly: 'How mad and bad she became at the end when she had lost her beauty and was losing her wits. And what a bitchy old bore she was, telling the same scandalous obscene stories over and over again' (*TSTS* 68). However, on meeting his childhood sweetheart Hartley quite by chance in the village close to Shruff End, it is she who becomes the 'alpha and omega' of his 'sacred tale' (*TSTS* 78). He remembers that her 'body was passive to my embraces, but her spirit', so he claims, 'glowed to me with a cold fire' (*TSTS* 80). Charles also reveals that, years ago, his teacher Mr McDowell would not tell him of Hartley's whereabouts, fearing 'I might commit some act of violence' in the 'raging bitterness of jealousy' (*TSTS* 83). He attributes his countless subsequent love affairs to Hartley's rejection of him. Charles affords himself the satisfaction of imagining, in another authorial nod to Valéry, 'that she felt terrible secret pangs of remorse and regret, and that a bitter worm gnawed her as painfully as it had gnawed me' (*TSTS* 85): '*Le vrai rongeur, le ver irréfutable / N'est point pour vous qui dormez sous la table, / Il vit de vie, il ne me quitte pas!*', translating as, 'The real, the irrefutable gnawing worm / Is not for you, asleep beneath the table, / He lives on life, he never gives me rest!' (Valéry 1971, 18–19). The village church with seaside cemetery becomes the couple's meeting place and while he awaits her, Charles lies back upon a 'lichen-spotted' gravestone 'sketching a kind of programme for survival' (*TSTS* 120–21). He believes that what he 'must now concentrate upon was the possibility of love in the form of a pure deep affectionate mutual respect, a steady constant binding awareness' (*TSTS* 121). This is pure fantasy from Charles.

4 The Sovereignty of the Sea in *The Sea, The Sea*

He had hoped that coming away to the sea would render him 'pure in heart', but his 'aspirations to sanctity' conflict with an 'urgent dark desire to *act*' (*TSTS* 122, 138). Charles's predilection for cruel behaviour precludes administering proper attention to Hartley. In fact, despite Hartley's deep affliction, he derives a spiteful pleasure from learning that her son Titus is missing. Charles's subsequent confrontation with Hartley's husband, Ben, is at best ambiguous. Charles presents Ben as a 'hateful tyrant', despite having to admit the need to calm his own 'violent feelings' and block 'certain terrible sensations' because Ben chooses not to know Charles (*TSTS* 152). A 'frightful crippling mindless jealousy-pain' comes over Charles after the 'horrible interview' which makes him feel 'free to detest [Ben]; and [...] free to do more, oh ever so much more than that' (*TSTS* 157). In a 'dreadful violent leaping ahead', Charles becomes set on rescuing Hartley from what he perceives is her 'lifetime of unhappiness' (*TSTS* 157–58).

Violence underpins the conversation between Charles and his long-standing friend Peregrine who has never heard mention of Hartley before this moment. In a drunken conversation, Charles casually asks him if he hits his women. For Peregrine, such an idea represents the 'last barrier of civilisation' (*TSTS* 166). Nonetheless, Charles wants 'increasingly, and with a violence which almost burnt the tenderness away, to own [Hartley], to possess her body and soul' (*TSTS* 186). Later, eavesdropping on Hartley and Ben at home, Charles gets 'what he came for' and insists that what was said is 'written [...] out as I remember it', at once claiming not to record, for his suggestible reader, 'the tones of voice, his strident shout, her whining tearful apologies' (*TSTS* 199) even as he does so, by implication. In telling the story of Titus, Charles persuades his reader of Ben's propensity to violence. This claim is soon destabilized, however, by Charles's own aggression in his attempts to restrain Hartley at Shruff End, where he finds it 'too hard to attempt to crush this pinching, kicking animal into submission' (*TSTS* 232). Ben's calm reaction to Hartley going missing would seem to suggest that the earlier heated dialogue between the married couple, reported by Charles, is a concocted fantasy on Charles's part (*TSTS* 236). Charles enlists Gilbert to stay with him in case Ben appears at Shruff End, because he does not know what might happen: 'Nothing violent, I hope?' entreats Gilbert (*TSTS* 270). Later when Charles returns Hartley home and Ben bids Charles good night, there is no sign of the husband's alleged violence. Charles insists Ben is 'a foul insanely jealous bullying maniac' even though Hartley denies being unhappy in her marriage (*TSTS* 307). 'But what makes you think my marriage is so bad, how can you judge?' she asks (*TSTS* 306). Charles appears at his most violent while holding Hartley captive: 'I felt I wanted to silence her even if it meant killing her' (*TSTS* 306). It is James who eventually persuades his cousin to free Hartley from her incarceration. Charles accepts that 'I ought never to have locked her up, I saw that now', but sinisterly qualifies this remark, 'I could easily have kept her, for a short time by strong persuasions' (*TSTS* 357). Ben will not 'always batter her mind and supervise her body' (*TSTS* 357) now she is home, Charles tells us, which evidently has been the nature of his own domineering behaviour towards Hartley.

When Charles declares Ben to be his assailant at Minn's Cauldron, Titus's reaction to such an idea is illuminating: 'I can't imagine him doing that, It's not in character, it's most unlikely' (*TSTS* 379). Yet when Titus later drowns, Charles's immediate reaction is the desire for vengeance against Ben: 'further killing was the only consolation; and, as it seemed to me then, to survive the murder of Titus I had to become a terrorist' (*TSTS* 391). Remorse only comes later. Whatever the cause of Charles's volatile behaviour, its effects are brought to bear on those people and things around him. His predilection for violence and domineering behaviour in his human relationships correlate with his desire to assert sovereign power over his environment. But the edge zone between land and sea, the intertidal zone of the shore is, however, capable of being at once both benign and perilous and it is before the neutral and unyielding sea that Charles experiences not only desperate tragedy, but brief moments of clarity and enlightenment.

The ocean does not provide the easy rest, or succour, that Charles presumes to receive from it. While he claims a 'sudden ecstasy' when Titus joins him in the water, the sea constitutes a further layer of confinement, 'an image of inaccessible freedom' (*TSTS* 263) for Hartley who, we are told, is unable to swim. The local people fear the ocean's power and it is therefore significant that Charles seems to have to record a significant tract of his so-called memoir back in London, away from the unassailable sea. When Charles does return, he survives a perilous shove into Minn's Cauldron. Later, Titus drowns, taken by the 'ruthless unchilding sea' (*TSTS* 459) as Charles would have it. Charles expresses bitter remorse for not wishing 'by mean prudence' (*TSTS* 402) to spoil the moment and forewarn Titus of the dangers. By contrast, the mystic James, displaying a deference to the mighty power of the sea, regrets: 'I never, for Titus, *watched* the sea' (*TSTS* 402). Murdoch's purpose here is to remind us of the perpetual elemental might of the ocean and, with it, the antithetically evanescent nature of human life.

'How Can We Individual Specks be Free and Moral?'

Murdoch's life's work can be generalized as a tireless effort to represent a world too complex to apprehend. In fact, this novel arguably serves as Murdoch's single most important treatise to the planet: its organic power, the fleeting nature of human life and our insignificant presence upon it. Early in the novel, Charles clambers out into darkness and onto the rocks 'very close to the sea and just above it' where he decides he will sleep: 'I could lie with my head on a cushion, looking straight out at the horizon [...] More stars were coming, more, more' (*TSTS* 144). Communing with the stars, Charles delusions of 'lifelong faithful remembering' transmute into a realisation 'as if the spirit that I prayed to had admonished me in reply', and he tries with some difficulty to put himself 'out of the picture and to pray only for Hartley', that she could be happy and that Titus might come home (*TSTS* 144). Yet, next morning he has become Friedrich's wanderer, once more seeing himself 'as a dark figure in the midst of this empty awful silent dawn' (*TSTS* 146). The starry night had momentarily offered him a sublime sense of his own insignificance. At dawn, he resists the desperate sense of loneliness to reassert himself back into his own picture of the seascape idyll and to continue in his pursuit of Hartley.

Later in the novel, having rediscovered his hastily penned account of his rescue by James, Charles decides to sleep out under the stars again in a scene that calls Keats to mind:

> —then on the shore
> Of the wide world I stand alone, and think
> Till love and fame to nothingness do sink. (2007, 100)

In the novel, this is the sublimely moral moment on the shore, listening to the sea and contemplating the stars, when it finally dawns on Charles that his 'vanity had killed Titus' (*TSTS* 471):

> As I lay there, listening to the soft slap of the sea, and thinking these sad and strange thoughts more and more and more stars had gathered [...] And far far away in that ocean of gold, stars were silently shooting and falling and finding their fates [...] And curtain after curtain of gauze was quietly removed, and I saw stars behind stars behind stars [...] And I saw into the vast soft interior of the universe. (*TSTS* 475)

Whereas his previous commune with the stars had induced a profound fear as a 'necessary captive spectator' (*TSTS* 146), it now dawns on Charles that his cousin saving him was a more significant act than he had realized. 'In sublime experiences we suffer and exult in the contest of Reason with the terrible contingent sinful world', explains Murdoch, these 'are great Romantic ideas which we may trace in many of Kant's successors in the existentialist line—Schopenhauer, Nietzsche, Sartre and Simone Weil' (*MGM* 133). She observes that wondering 'at the universe occasions a special (sublime) moral feeling: how can we individual specks be free and moral?' (*MGM* 443) Murdoch believed humanity to be capable of this moral leap: 'We can', she answers but, with notable circularity, warns that this 'is also the road to Romanticism' (*MGM* 443). Charles asks: 'Could I then learn to love uselessly and unpossessively and would this prove to be the monastic mysticism which I had hoped to attain when I came away to the sea? (*TSTS* 460). The Murdochian world is not, however, a neat and tidy one. Charles leaves the coast for London, and we never learn whether or not he retreats back into his domineering and coercive self.

'Conservation of Places and Waters, of Sensation and Recollection'

The 1970s was a hopeful decade, and *The Sea, The Sea*'s publication in 1978 'coincided with a prevailing spiritual awakening' and, more generally, a growing interest in Buddhism, according to John Burnside (1999, ix). In her adherence to her moral philosophy of the Good in place of a personal God, Murdoch certainly felt Buddhism had a great deal to teach Christianity.[12] The publication of *The Sea, The Sea* also coincided with the launch by Greenpeace, that same year, of its first Rainbow Warrior vessel—further evidence, should we need it, of the emerging ecology movement (Adams 2011). Environmental concerns were moving to the heart of society and becoming a political issue.[13] Murdoch had supported such concerns since childhood, joining the 'earliest' conservation group, Men of the Trees, which would become the International Tree Foundation (*IMAL* 43).[14] In *The Sea, The Sea,* Murdoch gives voice to conservation issues in the character of Titus. He is concerned for the 'preservation of whales', stands 'against pollution', and thinks the problem of nuclear waste is *terrible*' (*TSTS* 253). Featuring adherents to vegetarianism in many of her novels, Charles with his infamous eating habits is not one of those.

Murdoch was decrying the proliferation of plastic in our waters and on our planet over a quarter of a century ago. In her review of a cult book for *The New York Review of Books* about literary swimming enthusiasts, *Haunts of the Black Masseur: The Swimmer as Hero*, she reported:

---

[12] In his memoir, Bayley connects Murdoch's animistic sensibility with her respect for Buddhist beliefs when he writes, 'Iris's own private devotion to things finds a response in some of the tenets of Buddhism' (Bayley 1999, 120).

[13] President Richard Nixon heralded the 1970s 'the environmental decade' in a speech in 1970, see John Barr, 'Environment Lobby' (1970, 210).

[14] Murdoch penned the foreword to a memoir published in 1967 about Wendy Campbell-Purdie's afforestation initiative in North Africa, titled *Woman Against the Desert*. Murdoch writes: 'The importance and value of this project itself is obvious, and it is not my task to comment upon this. Other things in the tale seem to me to be noteworthy. This is a case of a good deed in a naughty world which has been brought about by the sheer creative energy of a single individual, and this at a time when we are getting wearily used to the idea that there is nothing more which single individuals can do' (Campbell-Purdie 1967, 12).

Today the tranquil Sargasso Sea is said to be full of plastic shopping bags, drifting indestructibly forever. I believe a submarine camera even spotted one hooked on to the deck of the sunken German battleship Bismarck, four miles down in an Atlantic deep.[15]

It is here that Murdoch connects the act of swimming to 'conservation—of places and waters, of sensation and recollection'.[16] She was conscious of the beneficial effect of tidal waters on the human mind; her thinking aligned with ecofeminist and philosopher Val Plumwood on 'the contemplation of the agency, power and mystery of places' (2008). Bayley's memoir, controversially published before Murdoch died, describes contributions he claims to have made to her writing; here, he records that it was he and not Murdoch who reviewed *Haunts of the Black Masseur*, writing 'under her name' (1999, 128).[17] Murdoch was obviously too unwell by this time to be able to offer nuance or context to Bayley's assertions.

Throughout her life, Murdoch was an avid swimmer of seas and of rivers, most notably the River Thames at Oxford. She connected the activity of swimming to the spiritual life, writing to Philippa Foot on 13 December 1989,

> I should be sorry to be suspected of approving of *God*. I believe heartily in his non-existence and indeed regard him as an obstacle to the spiritual life. [...] SWIMMING, that too is the thing, a very spiritual activity. (*LoP* 556)

She may not have wanted to be thought of as adhering to theistic belief but she was, nonetheless, admitting to something amounting to spiritual experience in immersing herself in the fresh river water.

---

[15] The cult book was Charles Sprawson (1993) *Haunts of the Black Masseur: The Swimmer as Hero*; it was reviewed by Murdoch (1993) 'Taking the Plunge', in *The New York Review of Books*.

[16] Murdoch (1993) 'Taking the Plunge'; for a discussion of Murdoch's own passion for swimming in wild places, *Podcast*, 23 August 2020.

[17] Bayley also relates in his memoir how Murdoch asked for his help with 'technical detail' such as how automatic pistols work, about cars and wine, for ideas relating to Charles Arrowby's eccentric eating habits, and for his assistance in devising the back story to Dora and Paul Greenfield's relationship in *The Bell*: 'the results are on page ten of the novel as first printed', he claims (Bayley 1999, 45–46).

4 The Sovereignty of the Sea in *The Sea, The Sea*

Murdoch was deeply concerned about the power of littoral spaces, and our human impact upon them. Murdoch shared, in interview, her deep respect for the ineffable power of tidal waters. She told John Haffenden, 'I used to think the sea and I were great friends, but one must fear the sea' (*TCHF* 127). In *The Sea, The Sea*, the seascape is drawn both to function independently and to intrude on proceedings:

> The rain came down, straight and silvery, like a punishment of steel rods. It clattered on to the house and onto the rocks and pitted the sea. [...] The flashes of lightning joined into long illuminations which made the grass a lurid green, the rocks a blazing ochre yellow [...] Tension and excitement and a kind of fear filled the house, the aftermath of mishap now somehow enacted by the elements. (*TSTS* 371–72)

The natural elements are seen to work in dialogue with her narrator-protagonist to play a critical part in this representation. As Bill McKibben expresses it, 'Nature's independence is its meaning; without it there is nothing but us' (1990, 54). Murdoch's representation of ecosphere is fundamental to the narrative.

It is clear that Murdoch engaged intellectually with the complexity of human reliance on the natural world, and, in turn, possessed a deep sense of planetary indifference to human affairs. It is, therefore, crucially significant that Murdoch chooses to foreground the seascape of the novel before introducing the novel's narrator. *The Sea, The Sea* seeks to picture the independence, the indifference and the constancy of the sea in contrast to the human workings, and as such provides a powerful if fleeting opportunity for revelation for Charles. By foregrounding the wild coastal environment, Murdoch is putting up a mirror to the frighteningly contingent nature of human life, subject to chance and mortality, in juxtaposition to a narrator-protagonist who, in denial for much of the story, fails to observe or interpret his surroundings. The authentically drawn richly biotic coastal environment juxtaposes and resists the sovereign demands of this self-obsessed and violent man. It is by paying attention to his natural environment that Charles comes to realize that he has been 'reading [his] own dream-text and not looking at the reality' (*TSTS* 499). Murdoch's project was one of human imperfectability. In her deep cogitations about the Good, attention to the natural world was one important means by which she felt one could relinquish the human tendency to solipsism, even if momentarily, and in a wider sense, dispel human delusions of significance, apartness, and primacy.

References

Primary Sources: Novels, Philosophical Writings

Murdoch, Iris. 1997a. 'The Novelist as Metaphysician' (NM). In Iris Murdoch, *Existentialists and Mystics: Writings on Philosophy and Literature*, ed. Peter J. Conradi, 101–107. London: Penguin.
———. 1997b. 'The Sovereignty of Good over other Concepts' (SGC). In Iris Murdoch, *Existentialists and Mystics: Writings on Philosophy and Literature*, ed. Peter J. Conradi, 363–385. London: Penguin
———. 1999. *The Sea, The Sea (TSTS)* (1978). London: Vintage
———. 2000a. *The Unicorn (TU)* (1963). London: Vintage
———. 2000b. *The Nice and the Good (NG)* (1968). London: Vintage
———. 2001. *Nuns and Soldiers (NS)* (1980). London: Vintage
———. 2003. *Metaphysics as a Guide to Morals (MGM)*. London: Vintage.

Primary Sources: Other Writings

Murdoch, Iris. 1993. Taking the Plunge. *The New York Review of Books*, 4 March 1993. https://www.nybooks.com/articles/1993/03/04/taking-the-plunge/. Accessed 10 January 2018.

Secondary Sources: Artworks, Books, Chapters, Journal Essays, Podcasts, Poetry, Newspapers, Sound Recordings, Webinars and Websites

Adams, Tim. 2011. Return of the Rainbow Warrior. *Guardian*, 11 June 2011. https://www.theguardian.com/theobserver/2011/jun/12/rainbow-warrior-greenpeace-anniversary-ethical. Accessed 15 August 2018.
Barr, John. 1970. Environment Lobby. *New Society* 384, 5 February 1970.

Bayley, John. 1957. *The Romantic Survival*. London: Constable.
———. 1999. *Iris: A Memoir* (1998). London: Abacus.
Blake, William. 2009. *Songs of Innocence and Experience*. London: Arcturus.
Bove, Cheryl K. 1993. *Understanding Iris Murdoch*. Columbia SC: University of South Carolina Press.
Buell, Lawrence. 1995. *The Environmental Imagination: Thoreau, Nature Writing, and the Formation of American Culture*. Cambridge MA: Harvard University Press.
Burnside, John. 1999. Introduction. In Iris Murdoch (1999) *The Sea, The Sea* (*TSTS*) (1978). London: Vintage
Campbell-Purdie, Wendy, and Fenner Brockway. 1967. *Woman Against the Desert: Tree-planting in the Sahara*. London: Victor Gollancz.
Carson, Rachel. 2015. *The Edge of the Sea* (1955). London: Unicorn Press
Clark, Timothy. 2012. Derangements of Scale. In *Telemorphosis: Essays in Critical Climate Change*, eds. Tom Cohen and Henry Sussman, vol. 1. Ann Arbor, MI: Open Humanities Press.
Cohen, Margaret. 2010. *The Novel and the Sea*. Princeton: Princeton University Press.
Coleridge, Samuel Taylor. 1997. *The Complete Poems*, ed. William Keach. London: Penguin.
Conradi, Peter J. 2001. *Iris Murdoch: A Life* (*IMAL*). London: HarperCollins
———. 2001. *The Saint and the Artist* (*S&A*). London: HarperCollins
Coupe, Laurence, ed. 2000. *The Green Studies Reader: From Romanticism to Ecocriticism*. Abingdon: Routledge.
Dipple, Elizabeth. 1982. *Iris Murdoch: Work for the Spirit*. London: Methuen & Co.
Easterlin, Nancy. 2012. *A Biocultural Approach to Literary Theory and Interpretation*. Baltimore: Johns Hopkins University Press.
Friedrich, Caspar David. 1809. *Der Mönch am Meer* [*Monk by the Sea*] oil on canvas, 110 × 1715 mm, Nationalgalerie, Staatliche Museen zu Berlin, Berlin.
———. 1817. *Der Wanderer über dem Nebelmeer* [*The Wanderer above the Sea of Fog*] oil on canvas, 980 × 740 mm, Kunsthalle Hamburg.
Guthrie, W. K. C. 2012. *The Greek Philosophers from Thales to Aristotle*. London: Routledge.
Heise, Ursula K. 2008. *Sense of Place and Sense of Planet: The Environmental Imagination of the Global*. Oxford: Oxford University Press.
Horner, Avril, and Anne Rowe, eds. 2015. *Living on Paper: Letters from Iris Murdoch 1934–1995* (*LoP*). London: Chatto & Windus.

James, Erin. 2015. *The Storyworld Accord: Econarratology and Postcolonial Narratives*. Lincoln NE: University of Nebraska Press.

James, Erin, and Eric Morel. 2018. Ecocriticism and Narrative Theory: An Introduction. *English Studies* 99 (4): 355–365. https://doi.org/10.1080/0013838X.2018.1465255.

Johns-Putra, Adeline. 2019a. Climate and History in the Anthropocene: Realist Narrative and the Framing of Time. In *Climate and Literature*, ed. Adeline Johns-Putra. Cambridge: Cambridge University Press.

———. 2019b. "We Have Lost Yardsticks by Which to Measure": Arendtian Ethics and the Narration of Scale in the Anthropocene. Keynote at the *Association for the Study of Literature and the Environment–UK and Ireland (ASLE-UKI) Biennial Conference*, 2019, 4–6 September 2019.

———. 2021. EASLCE 18th Webinar: 'Realism(s) in the Anthropocene: Representational and Ethical Challenges', 16 April 2021, https://www.easlce.eu/category/news/conferences/webinars/. Accessed 16 April 2021.

Jones, R.T. 1993. Introduction. In Daniel Defoe, *Moll Flanders*. Ware, Wordsworth Editions.

Kern, Robert. 2000. Ecocriticism: What Is It Good For? *ISLE: Interdisciplinary Studies in Literature and Environment* 7 (1): 9–32.

Leeson, Miles, Natasha Alden, Hannah Marije Altorf, and Lucy Oulton. 2020. Iris Murdoch and Swimming. *Iris Murdoch Podcast*, 23 August 2020.

Lipscomb, Benjamin J. B. 2018. *The Women Are Up to Something: Murdoch, Anscombe, Foot and Midgley and their place in twentieth-century ethics*. London Lecture Series. The Royal Institute of Philosophy, 19 October 2018. https://www.youtube.com/watch?v=EZZ0DnoupeY#action=share. Accessed 23 October 2018.

Lipscomb, Benjamin J. B. 2022. *The Women Are Up to Something*. Oxford: Oxford University Press.

Macfarlane, Robert. 2003. *Mountains of the Mind: A History of a Fascination*. London: Granta.

Mahon, Derek. 2005. 'The Seaside Cemetery' (2001). In *Harbour Lights*. Oldcastle: The Gallery Press ebook

McGrath, Hugh P., and Michael Comenetz. 2013. *Valéry's Graveyard: Le Cimetière marin translated, described, and peopled*. New York: Peter Lang.

McKibben, Bill. 1990. *The End of Nature*. Harmondsworth: Penguin.

Mentz, Steven. 2009. Toward a Blue Cultural Studies: The Sea, Maritime Culture, and Early Modern English Literature. *Literature Compass* 6: 997–1013.

Mitchell, W. J. T. 2002. Imperial Landscape. In *Landscape and Power*, ed. W. J. T. Mitchell, 2nd ed. Chicago: University of Chicago Press.

Moden, Rebecca. 2023. *Iris Murdoch and Harry Weinberger: Imaginations and Images*. Iris Murdoch Today. London: Palgrave Macmillan.

Mullan, John. 2006. *How Novels Work*. Oxford: Oxford University Press.

Plumwood, Val. 2008. Shadow Places and the Politics of Dwelling. *Australian Humanities Review* 44: http://press-files.anu.edu.au/downloads/press/p38451/pdf/eco02.pdf. Accessed 25 March 2018.

Prettejohn, Elizabeth. 2005. *Beauty & Art 1750–2000*. Oxford: Oxford University Press.

Price, Simon. 2003. Iris Murdoch: An Interview with Simon Price. In *From a Tiny Corner in the House of Fiction: Conversations with Iris Murdoch (TCHF)*, ed. Gillian Dooley, 148–154. Columbia SC: University of South Carolina Press.

Raban, Jonathan, ed. 1993. *The Oxford Book of the Sea*. New York: Oxford University Press.

Rowe, Anne. 2019. *Iris Murdoch*, Writers and Their Work. Liverpool: Liverpool University Press.

Sekula, Allan. 2002. *Fish Story*. Düsseldorf: Richter.

Sprawson, Charles. 1993. *Haunts of the Black Masseur: The Swimmer as Hero*. London: Pantheon.

Valéry, Paul. 1971. *Le cimetière marin*, trans. Graham Dunstan Martin. Austin, TX: University of Texas Press.

Wordsworth, William. 1965. I Wandered Lonely as a Cloud. In *The Poetry of Wordsworth*, ed. T. Crehan. London: University of London Press.

———. 2007a. The Wanderer. Book I of *The Excursion* (1814, eds. Bushell, Butler, Jaye et al. London: Cornell University Press.

———. 2007b. *Lyrical Ballads,* ed. Michael Mason. New York: Pearson Longman.

5

Maurice Merleau-Ponty, Embodied Mind and Vegetal Agency: *The Good Apprentice*

'THINGS EXIST AS PROLONGINGS OF <u>MY</u> BODY—BUT I THINK [OF] THEM AS OBJECTS IMPERSONALLY'

The Sea, The Sea's abandoned harbour, with its adjacent '"ladies' bathing place" […] so overgrown with slippery brown weed and so strewn with boulders tossed in by the sea' (*TSTS* 12), is emblematic of other neglected locations in the novels. In such spaces, Murdoch depicts the vegetal on the creep. Here, vegetation reclaims space once colonized by humans. Such activity signals the agential power of animate Earth; it is a power that perpetuates and reverberates from the deep past and into a long future. The contemporary concern, to set in balance human development and cultivation with the pressing need for the rewilding of land, may seem a world away from what are traditionally considered to be the central drivers of Murdoch's engagement with fiction. This is until one observes spaces, in certain of the novels, that are shown to have the power to transform her characters and, in turn, her reader. These spaces harbour the sort of transformations that speak to the novelist's sense of agentive power in the phenomenal world. Murdoch's work indicates a commitment to agency as a shared concept, emanating from multifarious sources.

© The Author(s), under exclusive license to Springer Nature Switzerland AG 2025
L. Oulton, *Iris Murdoch's Wild Imagination*, Iris Murdoch Today,
https://doi.org/10.1007/978-3-031-87833-6_5

Her descriptions of characters physically interacting with their environment confirm her conception of our human situatedness in an animate world. Her ideas anticipate elements of contemporary environmental philosophy and ecocritical thinking.

When Murdoch immerses a character in natural landscape in a heightened mode of affective engagement, her reader is able to share in a moment of transformative awakening. Murdoch portrays occasions where place and body combine to manifest as a collision of forces: body and world 'intra-act' (Barad 2007, 141). The term 'intra-action' emerges from contemporary feminist theorist Karen Barad's theory of agential realism wherein the world, possessed of a dynamism, produces a flow of agency between and around matter, things and people. Barad explains that the 'primary ontological unit is not independent objects with inherent boundaries and properties but rather *phenomena*' (2007, 139). They assert that, 'through specific actions, a differential sense of being is enacted in the ongoing ebb and flow of agency. […] [T]hrough specific intra-actions phenomena come to matter—in both senses of the word' (Barad 2007, 140). A form of discursive transaction appears to take place that reflects Barad's agential reality, where the focus resides not within a subject-object dichotomy, but in what they call an inseparable, material 'intra-action' (2007, 141). For Barad, agency is not an attribute, but rather 'an enactment' between (2007, 214). The ontological unit is not an independent thing/person/bird/ tree/stone; a kestrel does not exist without its habitat, nor, of course, can we or anything else exist without an environment. Instead, it lies in the 'congealing of agency', whose connection presents moments of 'intra-action', between setting and character, between place and person, between reader and lived experience (Barad 2007, 141).[1] In Murdoch's novels, characters respond to their environment, and the reciprocal impact of the collision of forces is seen arriving through the body to the character's consciousness.

As such, these depictions anticipate (by some years) recent affect theory that similarly rejects Cartesian separation of body from mind and, indeed, body from world. Affect, according to Melissa Gregg and Gregory

---

[1] Barad asserts that 'matter is substance in its intra-active becoming—not a thing but a doing, a congealing of agency' (2007, 141).

Seigworth, is 'found in those intensities that pass body to body [...] in those resonances that circulate about, between, and sometimes stick to bodies and worlds, *and* themselves' (2010, 1). For Weik von Mossner, 'all cognition is necessarily embodied and embedded, the body itself also becom[ing] a site of interpretation', and this interest that circulates 'between narratives, bodies, and environments', she says, 'is echoed within recent affect theory' (2017, 10–11).[2] Murdoch's moving portrayals of collisions of forces, emanating from character and from the environment, engender affect in her reader and draw attention to another potential site of interest in the current materialist strand of ecocriticism. Material ecocriticism also considers the interplay of human and non-human forces, enmeshing nature and culture to examine

> a distributive vision of agency, the emergent nature of the world's phenomena, the awareness that we inhabit a dimension crisscrossed by vibrant forces that hybridize human and nonhuman matters, and the persuasion that matter and meaning constitute the fabric of our storied world. (Iovino and Oppermann 2014, 5)

Aligning with Barad's theory of agentive reality, with its potential to democratize subjectivity through a 'distributive vision of agency' (Iovino and Oppermann 2014, 5), is the theoretical discourse of vegetal agency that I introduce into the Murdochian frame later in this discussion.

This chapter draws all these elements together to discuss Murdoch's prescient conceptions of both the embodied mind and vegetal agency, and culminates in a close reading of *The Good Apprentice*.[3] The formative years for Murdochian philosophy are crucial here, and examining Murdoch's personal correspondence and journals unearths some of the early influences that can be said to have inspired some of the most powerful moments of embodied affective interaction produced between the often solitary characters and her richly animistic landscapes emerging in the later fiction.

---

[2] For more on the growing alignment of affect theory and ecocriticism see Weik von Mossner (2017, 9–13).
[3] An earlier version of this close reading of *The Good Apprentice* was first published in *The Murdochian Mind* (Oulton 2022, 453–67).

Semantics Merchants in a Phenomenal World

Murdoch returned to Oxford in 1948 to take up a Fellowship at St Anne's College, following war work at the Treasury in London, a period working for the United Nations Relief and Rehabilitation Administration in London and Austria, and a year at Newnham College, Cambridge.[4] Her journal of the time (J3) not only records her thoughts, feelings and social interactions but, importantly, reveals that she was engaging in a range of philosophical ruminations and attempting to establish her own thinking. Like her friends and fellow philosophers Elizabeth Anscombe and Philippa Foot, Murdoch was frustrated by what she saw as the constraints and limitations of the mid-century logical-positivist orthodoxy that continued to dominate at Oxford at the time, with its core principle the necessary verification of any assertion through observation or experience.[5] In one of her first BBC radio broadcasts for the Third Programme, later published as 'The Novelist as Metaphysician' in *Existentialists and Mystics*, Murdoch deftly dismisses those inhabiting the linguistic philosophical landscape as 'semantics merchants' (NM 102). Their immutable position was also a reductive one as far as she was concerned; with logical positivism secure in its somewhat technical realm, metaphysics was to be almost entirely disregarded. Yet, Murdoch felt such a position could simply not be fit for purpose in a post-war, post-Holocaust, nuclear age: she was intent on establishing her counterargument.

On 17 March 1947, Murdoch's personal journal reveals a young philosopher preoccupied with a barrage of perplexing questions following a discussion with 'Pippa [Foot] about her conversation with Elizabeth [Anscombe] about the cogito' (J3 84). Murdoch is deliberating over an interrelated set of issues: 'what is the "feeling of understanding" apart from its many conceivable "verifications"?' (J3 85). 'But what about the unconscious constituents of "a thought?" (What is "a thought?")' (J3 86).

---

[4] For Murdoch's early philosophical career see Broackes (2012, 4–9).
[5] For the early professional relationship of Anscombe, Foot, Murdoch and Mary Midgley, see Lipscomb (2022, 22–49); and Clare Mac Cumhaill and Rachael Wiseman's *Metaphysical Animals* (*MA*) investigates how the four philosophers attempted to establish their thinking in response to some of the most unthinkable acts of the twentieth century that included the Holocaust and the detonation of atomic bombs over Hiroshima and Nagasaki.

As far as Murdoch is concerned, the prevailing determinism of the logical positivists failed, in particular, to take account of the 'felt realm' (J3 86): the role of the body and its senses. How might it be possible, then, to verify the role of the body and the senses? At the same time, Murdoch is alive to the complexities of an ongoing 'materialist/phenomenalist quarrel' (the degree to which mental states and neural states could be said to be synonymous) when she asks herself, 'Are material objects more real than sensa?' (J3 86). As counterpoints to these many and complex ideas that Murdoch interrogated are the sort of tasks that she knew she needed to address in order to progress her investigation. For example:

> Distinguish what we mean by [the] word 'understanding'—and what [the] feeling of understanding is. Consider the Q[uestion] re[garding] emotions, pains, colours etc. Link with [the] problem of perception. (J3 85)

The late 1940s was a profoundly formative period for Murdoch's emerging moral philosophy, and the interrogatory nature of these contemporaneous notes is testament to the complex and multifarious enquiry that was getting underway.

Central to Murdoch's investigation was her conviction that the logical positivists, in failing to account for the body's role in the cognitive process, were overlooking a fundamental constituent of thought: sensation. In its motor-sensory capacity, the body mediates perception of the external world's reality and, as such, the feeling body has to be acknowledged as centrally important to Murdoch. 'Concept of cause in phenomenal/material/physical worlds? Sensa are <u>sensations</u>—which implies a body', she reasons (J3 89). Murdoch notes the symbiosis of sense and thought that Latin is able to convey: *sensum* 'something sensed', derived from *sentio* 'to discern by the senses, perceive, feel, experience, think' (*OLD*). At the time, Murdoch was reading the work of scientist-turned-philosopher Charlie Dunbar Broad whose major interest was the mind-body problem.[6] Murdoch does not record which of Broad's works she was

---

[6] On her arrival at Newnham College, Cambridge, as a postgraduate, Murdoch's supervisor was C.D. Broad, the university's Knightsbridge Professor of Moral Philosophy; Murdoch told Queneau on 17 July 1947 of the arrangement, adding '*horresco referens* [I shudder to relate]': 'I'm sure [he] doesn't care for metaphysics, except the mathematical kind' (*LoP* 88).

studying, but her ideas seem likely to have been sourced from *The Mind and its Place in Nature*, first published in 1925, in which Broad states that

> we have to notice that there is in fact a most intimate relation between minds and living bodies. The minds that we know about are not disembodied spirits; they seem to be tied to organisms, to grow and decay with these, and to cease when these die. [...] Any theory of Reality which can claim to be even approximately adequate must take such apparent facts into account, and must contain a doctrine of matter and mind which shall be consistent with them. (Broad 1937, 9)[7]

In a footnote determining the association of the term 'moral psychology' with Murdoch (and Anscombe) Justin Broackes records that Broad also gave lectures on Moral Psychology at Cambridge, but what 'was new', Broackes explains,

> was the idea that philosophers [...] needed to raise as a topic for continuing debate the quite general picture of human beings, in their motivations and their cognitive and moral relations to their environment. That is the task to which Murdoch applied herself, using the term 'moral psychology'. (2012, 36n)

Broad's name, inscribed in bold letters at the margin of Murdoch's journal, appears alongside her brief notes relating to his sense-data theory, theory which Broad also addresses in his collection of lectures. Murdoch's records the following:

> Broad—somatic sense history, w[ith] outstanding sensa wh[ich] unite w[ith] it to form [a] general sense history.[8] Ontological & epistemological difficulties with S[ense] D[ata] [...] Observer's body is phys[ical] object (w[ith] special component, its somatic history). (J3 120)

---

[7] Broad's book brings together his Tarner Lectures, delivered at Trinity College, Cambridge, in 1923 (Purton 2007, 50).
[8] The term 'somatic' has Greek origins, σῶμα, relating to the physical body as distinct from the mind, soul or psyche (*OED*).

The 'special component' she refers to is the mind-dependent experience that Broad recognized as being fundamental to the cognitive process in his view of perception. Yet, Murdoch is absolutely clear that thought cannot simply manifest in a disembodied mind. She discerned, instead, a mind that awakens to its environment as it is situated in the body in the world; the world is felt or experienced through the body, through its senses. Later, she notes in her journal:

> Body sympathises w[ith] things—a mediation non-instrumental (as well as instrum[ental]!) I pass into [an] object something of my rel[ation] to my body. Sensation is immed[iate]—not [a] translation. Things exist as prolongings of my body—but I think [of] them as objects impersonally. (J3 155)

She senses unpremeditated interaction, the body in sympathy with the world. But what is to be understood by her use of the term 'prolongings' in such a mediation? Her meaning seems to fuse temporal extension with a sense of primordial longing, to convey an embodied sensory yearning that connects or interacts with the phenomenal world. She expresses sensory experience (both past and present) emerging into thought, arriving in the mind as a physical yearning.

It is clear that the sensory body reacting in advance of the mind to the materiality of the world is a conception that resonates with Murdoch. Towards the end of the twentieth century, neuroscience was starting to provide some of the physiological evidence to support such an idea. I have explained previously how '[n]euroscientists found that just as the mind influences bodily action, so the body's motor system influences cognition' (Oulton 2014, 8). It is now far more widely accepted that thought is 'profoundly and continuously informed by our embodied sensing', and the concept of embodiment works 'simply to embrace a more balanced view of our cognitive, indeed, human nature' (Clark 2008, 217). All this seems to have been intuitively understood by Murdoch early on in her philosophical career, as Patricia Waugh observes:

> In effect [Murdoch] foresaw the turn in psychology away from positivism and behaviourism and towards evolutionary theories and the rise of first

and later generations of cognitive neuroscience. These processes have culminated in our own contemporary accounts of the enactive, embodied and distributed mind and its function as a builder of inhabitable worlds. Central to this is the now scientifically established recognition of the centrality of emotion to our intuitive selection of the salient from the inexhaustibility of the choices of the real, and of the importance of attentiveness and empathy in our recognition of the needs and existence of others. (2012, 35)

For Murdoch, the body is the sensory force with which she is able to apprehend the world, and her conception not only anticipates, by decades, 'the "affective" turn' as Waugh proposes, but is central to an understanding of Murdoch's 'own theory of aesthetic and moral "attentiveness"' (2012, 35). Waugh suggests that Murdoch's approach 'might be cautiously welcomed as opening onto aesthetic domains of feeling, imagination, empathy, moral phenomenology and beauty' (2012, 35). Murdoch's conception of sensory body and enactive mind aligns with European philosophy and the French phenomenologists such as Jean-Paul Sartre and Maurice Merleau-Ponty.

'The Body Is Our General Means of Having a World'

We know from letters sent by Murdoch to Raymond Queneau that she was familiar with Merleau-Ponty's work: his lengthy monograph *Phénoménologie de la perception* [*Phenomenology of Perception*] was first published in 1945. On 31 March 1948, Murdoch writes to Queneau that she has been reading Merleau-Ponty's 'admirable *Phénoménologie*' and that she has been left with a conflicting sense of both 'admiration and confused disagreements' (*LoP* 106).[9] At the time, Queneau was a regular contributor to *Les Temps Modernes*, the distinctive post-war literary journal spearheaded by an illustrious editorial board comprising Simone de

---

[9] Murdoch had met Queneau in Innsbruck in 1946 and 'regarded [him] as her intellectual soulmate' (*LoP* xii); Merleau-Ponty's *Phénoménologie de la perception* was first published by Éditions Gallimard, Paris, in 1945.

Beauvoir, Merleau-Ponty and Sartre; this was until Merleau-Ponty and Sartre parted company acrimoniously over a profound political disagreement relating to the Korean War (*LoP* 618, 87n).[10]

Merleau-Ponty, in the line of Edmund Husserl and Martin Heidegger, adhered to the phenomenologist view that perception occurs through being a spatially and temporally embodied entity situated in place. For him,

> the body is eminently an expressive space. No sooner have I formed the desire to take hold of an object than already [...] my hand as that power for grasping rises up towards the object [...]. The body is our general means of having a world.
>
> (MM-P 2012, 147)

In *Phenomenology of Perception*, Merleau-Ponty deviates from the work of his existentialist and phenomenologist colleagues and investigates embodied situated experience; in his view, it represented a much neglected metaphysical concern relating to human existence in a phenomenal world. Later on, and uniquely among his peers, Merleau-Ponty came to reject the notion of human primacy over all life, in the conceptual shift to his ontology of 'wild being' (MM-P 1968, 116–17). On 16 October 1949, Murdoch discloses to Queneau her desire 'to meet M-P', wryly adding, '(if he is human)' (*LoP* 118). There is no evidence that the two philosophers ever did meet but what is very much in evidence is that Murdoch's engagement with Merleau-Ponty's work continued.

Writing to Queneau again in early 1950, Murdoch describes her two radio broadcasts for the BBC Third Programme as 'all rather Merleau-Pontyesque' (*LoP* 121–22).[11] She uses the first of these two talks on existentialism, 'The Novelist as Metaphysician' broadcast in March 1950, to define phenomenology for her British audience as 'an *a priori* theory of

---

[10] In Brussels after the war, Murdoch had been drawn to a café to listen to Sartre: 'His talk is ruthlessly gorgeously lucid—& I begin to like his ideas more & more' (*IMAL* 215–16); Merleau-Ponty attacks Sartre and his position on the Korean War in '*Sartre et Ultrabolshevisme*' (MM-P 1973, 185).

[11] The two programmes on existentialism were broadcast on the BBC Third Programme in March 1950. Edited transcripts were subsequently published as 'The Novelist as Metaphysician' and 'The Existentialist Hero' in *Existentialists & Mystics*, (NM 101–7) and (EH 108–15). For an account of how the BBC came to commission these programmes see Lipscomb (2022, 127–28).

meaning with a psychological flavour and a highly developed descriptive technique' which, she says, is 'key to the thought of Sartre and the others considered as novelists' (NM 102). In addition to Sartre, Murdoch lists Albert Camus and de Beauvoir as writers of 'phenomenological novel[s]' who also do philosophy and, here, she ordains Merleau-Ponty as their 'literary critic in chief' (NM 101).

She expresses her own particular interest, whereby phenomenology 'describes various general modes in which the subject grasps, sees or understands the world' (NM 103). And, here, she critiques 'one of Sartre's most remarkable books', *La Nausée* [*Nausea*], dismissing the sort of 'logical loneliness' of his Roquentin as a 'plunge into the absurd' (NM 107). Murdoch argues that if we attempt to bestow meaning upon the 'physical world, [...] in that we see it as the correlative of our needs and intentions, then this meaning could in principle vanish, leaving us face to face with a brute and nameless nature' (NM 107). She evidently senses danger in the Sartrean position, in what amounts to an imposition of an inward-looking self that attempts to find meaning in the outer physical world.

Murdoch's interest in the French existentialist was born of her 'Francophilia and a search for an alternative to the British philosophical tradition' (Broackes 2012, 18) and, indeed, her excitement at hearing the 'pop star' philosopher speak (*SRR* 1999, 10), rather than a particular affinity with the French philosopher's creed.[12] At about this time, Murdoch was beginning work on what would become her first published volume, *Sartre: Romantic Rationalist* (1953). Of Sartre, Murdoch wrote to her friend Leo Pliatzky on 30 October 1945, 'I don't make much yet of his phenomenology, but his theories on morals, which derive from Kierkegaard, seem to me first rate and just what English philosophy needs to have injected into its veins' (*LoP* 50).[13] Later, Murdoch tells Christopher Bigsby in interview that, although meeting Sartre in 1945 meant she had begun to think of herself as a philosopher, 'I don't think I ever was an existentialist. I think that my objections to existentialism went right back to my first meeting with it' (*TCHF* 98). Her quarrel with existentialism

---

[12] For an account of Murdoch's encounter with Sartre see *MA* 148–51.
[13] Leo Pliatzky was an admirer of Murdoch's at Oxford. They kept in touch through correspondence during the war and, later, he became a civil servant. He was able to provide the novelist with information on the inner workings of the civil service (see *LoP* 620).

is apparent in her second radio programme, 'The Existentialist Hero' (EH 108–15), broadcast on the Third Programme later in March 1950.

In 'The Existentialist Hero', Murdoch again critiques the person who 'tends to find nature absurd', where nature is presented as 'the brute and meaningless scene into the midst of which man is inexplicably cast' (EH 112). As far as Murdoch is concerned, the physical world exists in and for itself and she cannot accept the febrile repulsion experienced by Sartre's Roquentin: 'Alone, wordless, defenceless, [Things] surround me, under me, behind me, above me. They demand nothing, they don't impose themselves, they are there' (Sartre 1963, 180). Instead, as I suggest in *The Murdochian Mind*, Murdoch senses a physical world that 'commands its own presence and independence' (Oulton 2022, 460). To this end, Murdoch sustained her interest in the direction Merleau-Ponty was trying to take his distinctive brand of phenomenology, by collecting and (lightly) annotating English translations of Merleau-Ponty's *Signes* (1960) (*Signs* [1964]) and the posthumously published *Le Visible et l'invisible, suivi de notes de travail* (1964) (*The Visible and the Invisible: followed by Working Notes* [1968]).[14]

Merleau-Ponty connects a conception of time to a sense of our presence in the world. Time, he says, 'is not an object of our knowledge, but rather a dimension of our being' (MM-P 2012, 438). Time 'is literally the sense of our life, and like the world it is only accessible to the one who is situated in it and who joins with its direction' (MM-P 2012, 454). Horner and Rowe identify 'an echo of Merleau-Ponty's work' (*LoP* 618) in *The Nice and the Good* when towards the end of the novel, the widowed mother of Pierce, Mary Clothier, is suddenly overcome by the realization of a deep love for her long-standing friend, John Ducane. Murdoch renders an extraordinarily lyrical portrayal of the moment of Mary's embodied apprehension:

> something often so much less definite than pictures [...] ties our fugitive present to our past and future, composing the globe of consciousness. We

---

[14] Murdoch's copies of these works are in the Iris Murdoch Collections: Maurice Merleau-Ponty (1964) *Signs*, trans. by Richard McCleary. Evanston, IL, Northwestern University Press, IML1069; Maurice Merleau-Ponty (1968) *The Visible and the Invisible: followed by Working Notes*, trans. by Alphonso Lingis. Evanston, IL, Northwestern University Press, IML541.

think with our body, with its yearnings and its shrinkings and its ghostly walkings. Mary's whole body now, limp beneath the tall twisted acacia tree, became aware of John from head to foot in a new way. (*NG* 332)

The temporal echo is evident. In place of the 'prolongings' (J3 155), expressed by Murdoch in her journal entry some twenty years before, are Mary's embodied 'yearnings and [...] shrinkings and [...] ghostly walkings' (*NG* 332). Toiling in the garden on a baking hot summer's day and sheltering now in the shade of the acacia tree, Mary is overwhelmed by an embodied affective response, 'a physical convulsion like an electric shock' (*NG* 331), the sensory realization, in the lush surroundings, of this love arriving through her body into her consciousness. Murdoch offers up an extraordinary rendition of Merleau-Ponty's corporeity of perception in her character's somatic response: not abstract, not scientifically explained, but emotion engendered by being situated in the physical environment and experienced through the body.

The embodied physicality of inhabiting the world and its implications is explored by Sophie Grace Chappell in her philosophical enquiry into the nature of epiphany, *Epiphanies: An Ethics of Experience*. She examines how epiphany is experienced in relation to place, asking perhaps the most fundamental question in relation to the way in which our bodies and our minds might experience place: 'What is it like to be a human being?' (Chappell 2022, 188). She says:

> The relationship between my sense of my healthy body, and my sense of what I can do with my healthy body in the context and place that I find myself in, is intimate. The two are not identical; but they are essential to each other. My sense of the possibilities that are there for my body at present in its environment at present is constitutive of my sense both of what my body is, and of what that environment is. (Chappell 2022, 188)

Enactive self-awareness—our bodily awareness of place, argues Chappell, is fundamental to our experience of epiphanic moments.[15]

---

[15] In her book, Chappell quotes from Macfarlane's *The Wild Places* to convey how we risk losing the sense of our bodily awareness of place because of the way that many of us live now. Macfarlane says, 'We have come increasingly to forget that our minds are shaped by the bodily experience of being

Bate also explores the intimacy of body and place in his discussion of the Romantic poets in *The Song of the Earth*. He adopts the Byronic principle of '*Tuism*' (Bate 2000, 191) in which one is divested of egoism in love's intimacy, presented by Bate as countering the solipsism that is conventionally considered a characteristic of the Romantic sublime.[16] This erotic principle, Bate says, 'grants that we inhabit the world with our bodies' (2000, 191) and he alerts his reader to the clear alignment with Merleau-Ponty's thinking:

> Erotic perception [...] through one body [...] aims at another body, and takes place in the world, not in a consciousness [...]. There is an erotic 'comprehension' not of the order of understanding, since understanding subsumes an experience, once perceived, under some idea, while desire comprehends blindly by linking body to body. (Bate 2000, 191, cites MM-P 2012, 159)

In *The Nice and the Good*, in a pre-reflexive moment of Byronic *Tuism*, Murdoch renders the erotic image of Mary's body momentarily awakening to the knowledge of her love for Ducane. Murdoch connects erotic love to an embodied awakening of true vision and, in doing so, asserts its moral potential.

'The Flesh of the World'

When he published *Phenomenology*, Merleau-Ponty was criticized for retaining the ontological separation of passive object and perceiving subject. With his emphasis on the 'unitary consciousness of the perceiver'

---

in the world—its spaces, textures, sounds, smells and habits—as well as by genetic traits we inherit and ideologies we absorb. A constant and formidably defining exchange occurs between the physical forms of the world around us, and the cast of our inner world of imagination' (Chappell 2022, 186, cites Macfarlane 2007, 203). Chappell discusses Nan Shepherd's 'book of naturalism', *The Living Mountain*, a work written during the 1940s but not published until 1977; it is widely acclaimed by ecocritics. Chappell's interest lies in how Shepherd subjectively experiences and shares her sense of place: her intimate knowledge of the Wells of Dee, Aberdeenshire (Shepherd 2011).

[16] As Bate explains, in *Tuism*, the French second-person (familiar) pronoun '*tu*' is presented as the inverse of egoism and has come to be seen as the practice of putting another before oneself.

intact, he appeared to have retained the notion of 'the Cartesian *cogito*', according to Louise Westling, despite offering an account of 'the subject's immersion in the world' (2014, 33). In his first monograph, Merleau-Ponty asserts that a subject 'only achieves his ipseity [or sense of selfhood] by actually being in a body and by entering into the world through his body' (MM-P 2012, 431). He says:

> If qualities [or things] radiate a certain mode of existence around themselves, if they have a power to enchant, […] this is because the sensing subject does not posit them as objects, but sympathizes with them, makes them its own, and finds in them his momentary law. (MM-P 2012, 259)

Later, Merleau-Ponty would come alive to the 'consciousness-object distinction' he had retained in *Phenomenology* and this, according to Westling, is what he was setting out to rectify with *The Visible and the Invisible* (Westling 2014, 33, cites MM-P 1968, 200). In this unfinished work, published with the philosopher's working notes shortly after his death (he was 53), Merleau-Ponty was starting to develop an 'ontology in which individual beings are intertwined with the basic stuff, or flesh, of the universe, existing in a kind of reversibility with other beings and things' (Westling 2014, 33). Here, the philosopher articulates what he sees as the fundamental premise of his new work, that '[t]his environment of brute existence and essence is not something mysterious: we never quit it, we have no other environment' (MM-P 1968, 116–17). Merleau-Ponty evidently envisioned the participatory qualities of a natural world with humankind embedded at the interchanges of a broader ecosystem; Murdoch appears to convey this inhabited (past and present) participation of sensuous body and environment in Mary's 'yearnings and its shrinkings and its ghostly walkings' (*NG* 332). Today, Merleau-Ponty's ontology, expressed as a co-reliance of all living things, is central to the concerns of environmentalists and ecocritics in the twenty-first century, and *The Visible and Invisible* has drawn renewed attention in recent years, with Merleau-Ponty, himself, acknowledged as an ecophenomenologist.[17]

---

[17] For more on 'ecophenomenology' see Brown and Toadvine (2003, ix–xxi).

5 Maurice Merleau-Ponty, Embodied Mind and Vegetal Agency...

One can conceive of Barad's conception of the interchange of energy between agentive forces evolving from Merleau-Ponty's ontology.

If phenomenology is characterized as the study of perception with the concomitant sensory qualities of seeing, hearing, feeling and so on, informing bodily awareness and somatic response, what then of ecophenomenology? During the final decade of Merleau-Ponty's short life, his singular perspective underpinned a core belief that all living organisms 'exist intertwined and in constant interaction with the flesh of the world around them' (Westling 2014, 26–27). Our sensory bodies engage in perceiving the world and, given that other natural entities or beings perceive somatically too, our involvement is as an integral and reciprocal presence. While phenomenologists following in the wake of Husserl and Heidegger continued to assert a conception of human primacy over all other life forms, Merleau-Ponty's ontology delivered its own unique perspective—the 'coevolution of all living things and his ontology of "wild being"' (Goodbody and Rigby 2011, 8). He attributed the notion of 'chiasm' (χίασμα means diagonal crossing) to his thinking, presenting a mental picture of the entwining relationship of this constant interaction of organisms and world, that as Westling has suggested might rather resemble the helix of the DNA molecule, 'which is the wild or brute being in which we are immersed' (2014, 26–27). In the end, Merleau-Ponty's legacy was secured by his articulation of the world's fundamental interconnection and co-reliance of all living things. It is this conception that engages the interest of modern-day discourses on the environment, from the realms of habitat conservation to environmental justice. In the case of environmental justice, in particular, it is interesting to examine how Merleau-Ponty's embodied ontology might have a role to play in Murdoch's concept of attention.

'Uninhibited Descriptions'

Embodied sensory perception is crucial to Murdoch's concept of attention. Denham constructs an image of the Murdochian vision of moral perception where 'moral experience features a quasi-experiential phenomenology analogous to sensory (and particularly visual) perception'

(2012, 328). In Denham's view, this confirms the intrinsic contribution that emotion makes to moral judgement. Horner and Rowe's observation in *The Nice and the Good*, then, gestures towards a significant, but to date under-appreciated, formative influence of the French phenomenologist's work on Murdoch as philosopher and novelist. Merleau-Ponty's particular appeal may have been that, akin to Murdoch, he 'was dedicated to the task of describing experiences as closely and precisely as he could' (Bakewell 2016, 228). Far removed from the 'semantics merchants' of logical positivism, and in significant contrast to the 'free and lonely self' of Sartrean existentialism (NM 104), Merleau-Ponty sought out a form of compromise with the world. For him, the perception of being situated in the world we might intuit as children was key: moving through the world we meld curiosity and experience with our senses in order to apprehend it.

In *Metaphysics as a Guide to Morals*, Murdoch revisits her earlier assessment of Sartre and considers him in relation to Merleau-Ponty too. By this stage, she has become yet more dismissive of the Sartrean credo: the key image being *être-pour-soi* [being for itself], the free individual self, in contrast to *être-en-soi* [being in itself] the contingent surroundings, discussed in her early work on Sartre.[18] Now, she argues:

> What is pure and strong and free (*être-pour-soi*) emerges as it were *automatically* beyond and above what is dull, jumbled and senseless (*être-en-soi*). Thought moves bodiless and unimpeded above this morass, it makes choices and carries values. People who cannot think, who fail to think, who choose not to think, are pictured as being sunk in a dark muddled consciousness, composed of feelings and associations and fragmentary awareness. (*MGM* 156)

This key image, of 'bodiless' Sartrean thought, moving 'unimpeded above this morass, [making] choices and [carrying] values' (*MGM* 156), clearly contradicts the embodied mind that Murdoch so vividly intuits. However, by the time Murdoch comes to deliver her Gifford Lectures, and shortly before she sets to writing *The Good Apprentice*, Merleau-Ponty's brand of

---

[18] See Murdoch's 1987 introduction to *Sartre: Romantic Rationalist* (*SRR* 1999, 9–11).

5 Maurice Merleau-Ponty, Embodied Mind and Vegetal Agency...

phenomenology-in-the-flesh also appears to have lost its attraction as a plausibly defensible phenomenological approach. In one of her Gifford Lecture manuscripts she demurs,

> Changing attitudes to phenomenology may be traced in the career of Merleau-Ponty. Husserl's method tends (as it seems to me) to fall apart into abstract logical considerations and rather unclear description, which seem to belong to empirical psychology or even to the art of the novelist. 'Concepts without intuitions are empty, intuitions without concepts are blind': phenomenological analysis risks an inconclusive division along these lines. (Giffords 1982, 200)

Here, she summons Kant to reinforce her argument.[19] Later, in *Metaphysics as a Guide to Morals* she assesses phenomenology's shortcomings in similar vein: 'Phenomenological analysis risks an inconclusive division, falling apart into either abstract logical structuring or uninhibited descriptions which may seem to belong to empirical psychology or even to the art of the novelist' (*MGM* 158).[20] What seemed lacking in Merleau-Ponty's chiasm for Murdoch was ontological rigour. Entwined in the flesh of the world, how might one attend to the other? Indeed, how in this enmeshed circumstance would one even be able to define other?[21]

---

[19] Murdoch paraphrases Kant, 'Thoughts without content are empty, intuitions without concepts are blind', in *Critique of Pure Reason*, A51/B76.

[20] *Metaphysics as a Guide to Morals* includes much of the material from the Gifford Lectures (although *MGM* does not expressly say so); she makes this similar point, although there is more detail on Husserl as a '"bridge" figure' whose 'psychological intuition [...] lacks the precision of either science or philosophy' and she no longer paraphrases Kant (*MGM* 158).

[21] I am grateful to Silvia Caprioglio Panizza, Antony Fredriksson and all the participants of the online workshop 'Ethical Attention: Iris Murdoch in philosophical dialogue', 4–5 February 2022, for the rich discussion on this particular issue. The workshop was a joint initiative of the Centre for Ethics in Public Life, University College Dublin, and the Centre for Ethics, Univerzita Pardubice. Drawing on Merleau-Ponty's phenomenology, Fredriksson examines the implications for the ethics of attention of intersubjectivity—a shared world of embodied beings, human and non-human—in his book *A Phenomenology of Attention and the Unfamiliar: Encounters with the Unknown*. Fredriksson argues that our 'sense of the world is not solely constructed by [our] own devices', when, for example, it takes a dog reacting to a squirrel to cause Fredriksson to perceive the squirrel himself: 'Through the attention of the other, I may discover aspects of our common world that are partially hidden for me' (2022, 111). Fredriksson asserts, 'What I want to indicate here is that the representationalist framework for philosophy of perception fails to explain how we [he and his dog] are able to attend to the same thing without any mindreading, shared life-form, shared language, or theorizing and simulation. [...] [T]he action of guiding the other in seeing the same thing is [...]

Murdoch's notes inscribed on the endpaper of her own copy of Merleau-Ponty's *Signs* suggest a similarly sceptical view. She declares: 'Ramblings of this sort [are] only ok if illustrated in detail in discussion of lit[erary] work' and, she adds in parentheses, 'ie. lit[erary] crit[icism]'.[22] Whereas Murdoch struggled to clarify and defend her ideas on the embodied mind as a writer of philosophy, she displays no such inhibitions as a writer of novels. By the 1980s, Murdoch has largely dispensed with Merleau-Ponty's immersive ontology and yet (as might be inferred by the quotations above), she carries the substance of the ecophenomenologist's thought into the 'uninhibited descriptions' (*MGM* 158) of her further novel-writing. This is particularly in evidence in *The Good Apprentice*.

The Immersive Phenomenal World of Seegard

A discussion of the setting of Seegard, in *The Good Apprentice*, essentially calls for a few remarks on Murdoch's approach to realism. Murdoch tells W. K. Rose, circa 1968:

> I would like to be thought of as a realistic writer, in the sense in which good English novelists have been realists in the past. I want to talk about ordinary life and what things are like and people are like, and to create characters who are real, free characters. (*TCHF* 29)

Contrary to Murdoch's particular claim to be a writer of realism, her setting of Seegard assumes a mythic, almost folkloric, quality, and it is one that has invited a range of critical interpretations, some of which dismiss the notion of realism altogether. Of *The Good Apprentice*, Conradi remarks that

---

established through our common embodied orientation in the external perceivable world. In this understanding, perception is an action, an active engagement with the environment and other beings' (2022, 111–12).

[22] See Murdoch's copy of Maurice Merleau-Ponty (1964) *Signs*, IML1069, in the Iris Murdoch Collections.

Murdoch's joy in the creation of this separate world with its own charms of physical layout and of emotional logic, recalls her (and our) pleasure in Imber Court in *The Bell*, Gaze Castle in *The Unicorn* or the Rectory of *The Time of the Angels*, and it is the point about Seegard, like Imber, Gaze and the Rectory, that each represents some kind of fey spiritual halfway house, a middle realm gone wrong, aspiring and corrupted. (*S&A* 334)

Conradi focuses on the establishments themselves: the richly conceived exterior landscape that forms an intrinsic part of the Seegard location is not pictured in his assessment of Murdoch's 'separate world' (*S&A* 334).[23] Fiddes sees a Murdochian patterning in Seegard: a house with 'prisoners' (2022, 49) that is a feature of a number of the novels such as Hannah in *The Unicorn* and, less ambiguously so, Carel Fisher's daughter Muriel, his ailing ward, Elizabeth, and his maid and sometime mistress, Pattie, in *The Time of the Angels*. Like Conradi, Fiddes views the Seegard arrangement metaphorically: it represents the storied world we, as humans, tend to create around ourselves. Fiddes explains:

> Instead of giving *attention* to people and things as they really are, noticing their reality and through them attending to the Good, the greedy self imposes and projects itself on others and traps them within a web of its own devising. (2022, 49)

According to Rowe, Murdoch 'takes readers on a journey into a metaphorical underworld to explore the seductive attraction of evil and acknowledges the potential of art as a collaborator in its power' (2019, 32). Martin and Rowe point to a complex layering of imagery that includes the 'metaphorical implications of Seegard and the symbolism within Jesse's paintings [that] defines evil as the unregulated operation of base desires unmediated by any influences from the outside world' (2010, 141). While accounting for the mind (Edward Baltram's, specifically), such exegeses do not seem to fully account for Murdoch's depictions of the sensory bodies of her characters in the world of this novel, or her

---

[23] Conradi offers an eloquent assessment of the 'physical layout' of Seegard's interior that includes the 'parodic familiar' of daily routines that face up to 'strange noises', presenting a 'fake' vision of beauty functioning somewhere between 'holiness and magic' (*S&A* 334–35).

practice of conveying place by vicariously triggering the sensory perception of her reader. David Cooper does acknowledge Seegard's realist treatment. He describes it as an 'isolated, watery and faintly sinister retreat in the country' in his introduction to the novel (Cooper 2000, ix). However, more generally, the foregrounding of symbolic imagery by critics does not seem to account for the evident influence of the expansive countryside surrounding Seegard on Edward, in which he begins to immerse himself and from which he gradually begins to gain succour. There is no single prevailing critical view of Seegard and its implications, and it is interesting that Murdoch, herself, appears to take a protean view of the novel's interpretive potential. Take, for example, the multitude of interpretations she offers Richard Todd for the character of Edward's father, Jesse Baltram:

> I [...] think of Jesse as a supernatural figure. But of course you can also regard him in an ordinary way, as just a harmless, decrepit old man. Or you could indeed think of him as a sort of junior god, a kind of magician who comes back to earth at different times (perhaps there are many of them). In the story, his power has become exhausted, so perhaps he has to retire somewhere for three thousand years before he comes back again. (*TCHF* 185–86)

Murdoch reveals her consummate skill in the art of creative ambiguity, and its significance:

> You could look at it either way. Things may not be perceived as supernatural; one feels this about the presence of what looks like magic in the world, strange things or paranormal things. But lots of ordinary human relationships, too, may move into an area of mystery. (*TCHF* 186)

These represent the shape-shifting qualities of her novels that help sustain Murdoch's relevance, and readers' interest, into the twenty-first century.

Sabina Lovibond helpfully poses a more general question on realism: 'what does it take for a work of fiction to qualify as "realistic"?' (*IMR5* 36). In answer to her own enquiry, Lovibond refers us to Henry James who would suggest, she argues, that realism is more concerned with 'artistic success, irrespective of method' (*IMR*5 36). And, in recognizing Murdoch's admiration for James, Lovibond adopts the Jamesian 'solidity of specification' measure (that, he avers, helps to give 'life' to the novel), and she applies this measure to Murdoch's later fiction (*IMR*5 36, citing James 1948, 12).[24] Lovibond believes that

> The areas of maximum 'solidity' [...] are those of personal appearance or physiognomy, and of physical setting in general: the natural environment (sea-coasts, river-banks, meadows, overgrown places) and its flora and fauna; and the human, or social, environment in so far as this falls under the heading of 'real estate'—houses, gardens, large institutions. (*IMR*5 36)

Furthermore, I would add to Lovibond's remarks that readers of novels tend often and imperceptibly to use markers of their own individual experience to apprehend story. For example, in our readerly imagination, a tree in *The Good Apprentice* tends to become a tree from our own experience—except in that one particular instance, of course, where one might be taken in by 'that old crook' Mr Blinnet (*GA* 419) and the tree could indeed be Mrs Blinnet 'grow[ing] cold in her skirt of earth' (*GA* 389). But the serious point is that, as a reader, one tends to overlay often very personal experience of place onto a given literary space. Even if Seegard is read as symbolic of the unconscious mind or suggestive of a mythic space, how readers navigate that space remains inherently dependent on what just such a place (woodland, floodplain, mountain, seascape) signifies in a reader's own lived experience, and how that place makes them feel. So, while one can accept that Seegard does stray somewhat from the realism that Murdoch lays claim to, it takes one's individual experience as reader to notice this fact and to ground the setting in a reality of one's own. In this way, the landscape in the novel takes on significance. In *The Good Apprentice*, Murdoch portrays affective interactions with the 'flesh of the world' (MM-P 1968, 116–17), which confirms Murdoch's continuing engagement with Merleau-Ponty in her novel-writing, which, in turn, anticipates the interest in the French philosopher among contemporary ecocritics today, and indeed the work of philosophers such as Barad.

---

[24] Lovibond lists the following references as evidence of Murdoch's admiration for James, *TCHF* 80, 94 and 226.

'Damp Smells of Spring'

A deeply remorseful Edward Baltram is beset by temporal blankness, having administered drugs and induced the accidental death from an open window of his friend Mark Wilsden: 'One momentary act of folly and treachery had destroyed all his *time*' (*GA* 11). The narrow streets of Soho seem claustrophobic and Edward longs 'for sleep, unconsciousness, blackness, the absolute absence of light' (*GA* 10). The present has become fixed, immutable. His stepbrother, Stuart Cuno, seeks a form of blankness for himself, too, and an invisibility in the service of others. Nonetheless, he entreats Edward to attend to, and to hold onto, the sound of the birds singing as he wakes each morning: '[H]old onto that *after* you've remembered, and just think "the birds are singing", and hold that away from the blackness and keep it there' (*GA* 52).[25] But Edward rejects the intrusion from the world outside: 'Damp smells of spring, of wet earth and green things growing, which would have made him happy once, came through the window which he closed with a bang' (*GA* 52). Stuart's apparent absence of self extends to an absence of feeling. To him, deaths such as Mark's are 'sad static things' along 'the road into the whiteness of his own future' (*GA* 55). Stuart's preoccupation is with 'a world without religion, of crazed spirit without absolute [...] of decay of human language and the loss of the soul' (*GA* 58). He has no desire to immerse himself in the phenomenal world either.

Edward leaves for his father Jesse Baltram's house, the mysterious Seegard. The smell ('of human company and things not yet irrevocable' [*GA* 104]) of the bus that takes him to Seegard is Edward's last reminder of the city, though the countryside does not immediately fulfil his preconceptions. To the newly arrived Edward, the landscape is 'bland', 'flat', 'waterlogged' and 'muddy', with a 'large sky', 'brown clouds' and 'drab earth' (*GA* 104–5). He considers his rural surroundings to be contingent and unappealing. He is quite unable to find the 'charm' (*GA* 104) of nature

---

[25] Of sound in the novels, Gillian Dooley says it 'is not atmosphere for its own sake. It is a profound element in Murdoch's most important endeavour, creating and embodying her characters', whose internal preoccupations too often conflict with an 'openness to the beauty of the world' around them (Dooley 2022, 98).

he instinctively seeks. With his estranged father mysteriously absent, he joins his stepmother, Mother May, and his two half-sisters living in the highly structured Seegard 'city state' (*GA* 136). Edward appears unmoved by his surroundings.

After the first few days, and 'establishing some quite new sense of being alone', Edward dares to explore his environment and to venture out to the nearby woodland. In stark contrast to the repulsion expressed in *Nausea* by Roquentin, 'What [is] the use of so many trees which [are] identical?' (Sartre 1963, 190), Murdoch's landscape appears to pulsate with the ethereal presence of a variety of ancient woodland trees:

> The wood, clearly the work of nature not of man, was a wonderful mixture of every sort of tree. There were oaks and ashes and beeches and larches and firs and wild cherries and some of the largest yews Edward had ever seen. (*GA* 127)

She celebrates the splendour of arboreal diversity, elegantly conveying the 'complex connections [made] between psychological well-being and the environment' that Rowe observes in Murdoch's work (2019, 106). Edward arrives at a place:

> Of course the wood was full of places, celebrations and juxtapositions, mossy alcoves, primroses showing off in the dead bracken, circlets of greenery where the sun managed to shine, long fallen trees as clean as bones. (*GA* 128)

Edward encounters a glade or clearing with a large vertical stone at one end whose regularity suggests 'some work of human intelligence', somewhere both ancient and sacred (*GA* 128), presenting a picture that figures in the minds of some readers as a nod to Heidegger. Late Heideggerian thought favours something approximating the ecocentric conception of humans as 'participants in a wider clearing of Being' (*Dasein*), as opposed to his earlier more anthropocentric notion of the human as being the 'space or "clearing" itself in which things show up *as* things' (James 2018,

190).²⁶ Gary Browning avers that what Murdoch particularly admired in Heidegger's thought was 'his overcoming of dualisms and his capacity to provide a unified but differentiated perspective on experience' (2018, 16–17). Murdoch brings Edward into the light, the clearing representing an enmeshment of both material space and consciousness.

When Edward attempts to leave and find his way back through the undergrowth, he experiences an instance of profound revelation. Again, Murdoch creates the moment of affective embodied apprehension that precedes cognition. Like Mary's experience of 'physical convulsion like an electric shock' (*NG* 331) in *The Nice and the Good*, the solitary Edward now undergoes 'something like physical change' (*GA* 129). It is as if

> some intense infusion were blowing into his face and enveloping his body. His head seemed to be opening up into a vast area, as if it were literally painlessly splitting and being joined to some enormous pale cloudy sphere up above. Thoughts then came in a rush. (*GA* 129)

In the 'labyrinth of colonnades and archways and vaulted halls and domed chambers' (*GA* 127) of an ancient wood, Murdoch pictures Edward's sensate experience of epiphany arriving through the body before entering his consciousness. At this moment, rather than concealing his 'awful guilty loving mourning for Mark' and finding 'some relief in running away', Edward begins to see himself in a place of pilgrimage, the prodigal son returning to his father, 'carrying his woeful sin to a holy shrine and to a holy man' (*GA* 129). The richly diverse woodland close to his estranged father's home—at once contingent, strange and disorienting—offers Edward a new perspective on his awful suffering.

When, later that spring, Edward returns to the same clearing where the saplings are now in bud and the bluebells radiating 'a hazy blue distance' between them (*GA* 168), he glimpses his half-sister Ilona who 'looked like [she] was lifted from the ground by some superior force' (*GA* 169):

---

²⁶ Murdoch laboured over a planned book on Heidegger which she completed in 1993 but decided not to publish; see 'The Manuscript on Heidegger', KUAS6/5/1/4 from the Iris Murdoch Collections. The manuscript is due to be published as Iris Murdoch, *Heidegger: Pursuit of Being*, ed. by Justin Broackes (Oxford: Oxford University Press, forthcoming).

her swaying body was carried away along the glade and then back again toward the pillar. [...] [T]he purposeful grace of her body, the patterned weavings of her arms, and of her long slim legs under the hitched-up skirt were those of a dancer. [...] It was a dance of joy, becoming slower and sadder toward the end, as if she felt the breath failing which had lifted her. (*GA* 170)

Ilona's embodied 'ecstatic yet disciplined expression' (*GA* 170) contrasts with the demeanour of an 'awkward schoolgirl' (*GA* 171) with which she shortly attempts to leave this hallowed space. At the end of the novel, the transformation in her poise and composure will become starkly evident when Ilona, now living in the city, is found by Edward working as a stripper and quite unable to dance. Here, her 'pallid, clammy, bare, [...] human form [is] revealed in all its contingent absurdity' to a darkened room 'entirely filled with men' who, '[b]lasphemously simulating the selfless contemplation of the mysteries of art or religion, [sit] tensely still, while inside each head a small machine of secret repetitive fantasy noiselessly whirred' (*GA* 496–97). This seedy portrayal of men in a strip-joint represents a grim picture of fantasy for Murdoch. The 'contingent absurdity' of the darkened room inhibits Ilona, who once found corporeal inspiration in the lush embrace of the arboreal environment at Seegard.

Stuart, Edward's half-brother, undergoes a form of embodied epiphany too. Having decided 'not to enter the machine' but in need of a strange form of spiritual transcendence, he appears simply to have opted out and to lack any sort of engagement with or empathy for his stepbrother's awful predicament: 'Stuart is too self-obsessed. He scarcely knows Edward exists' (*GA* 27), observes Stuart's father, Harry, early on. Now, Stuart stands accused of bewitching Midge and inflicting permanent damage on her son, Meredith (*GA* 472). Standing on a station platform, it occurs to Stuart in a moment of 'shameful loneliness' that he 'was condemned for eternity to be a useless and detested witness of [the] sufferings' of the human race (*GA* 479). Instead of gazing blankly at the tracks as he was in the habit of doing, while filling his head with the usual messianic imaginings of the heroic deed he would perform should someone fall onto the line, he focuses in on something moving along, a mouse going about its ordinary existence: '*It lived there*' (*GA* 480).

> This revelation was taken in by Stuart in a moment. It entered him like a bullet. It exploded inside him. He felt about to fall. He stepped back from the edge of the platform. He found a seat and sat down, leaning his head against the tiled wall. What had happened, was he having some sort of fit? He gasped for breath, feeling his whole body change. An extraordinary peaceful joy ran through him, a thrilling consciousness of the warmth and pace of his blood, running through all his veins and arteries down to the minutest vibrating threads in his finger tips. A light shone in his eyes, not painful, not a flash, but like a shrouded sun which warmed his body until it glowed as if it too were all radiantly alight. He rolled his head to and fro against the tiles, half closing his eyes and sighing with joy. (*GA* 480)

This moment of enactive epiphany for Stuart is seen arriving through his body suddenly, in this instance, 'like a bullet' (*GA* 479). Despite the potential for the mouse to be seen as innocuous and the moment to seem as fleeting as the mouse itself, Stuart's epiphanic experience has profound impact. Critics have found the sequence to be wholly significant. Rob Hardy observes that 'Stuart's image of the mouse is the symbol of initiation produced by his unconscious—yet it is not his alone. The mouse is a living creature in the real world outside Stuart' (2010, 53). Conradi says that Stuart 'finds spiritual help from watching mice living on the Underground tracks', help that Stuart needs (*IMAL* 566). Martin and Rowe's interpretation of this important moment is rather more expansive:

> Stuart's epiphany at the centre of the novel as he watches a tiny mouse living deep underground among the tube rails, accepting that '*it lived there*' (*GA* 480), triggers his own understanding that he has to brave and participate in the entirety of human experience and confront the capacity for evil within himself before he can become a good man. (2010, 144)

Conceiving of the contingent life of an ordinary mouse beneath the rails brings life flooding back to Stuart. The experience of truly seeing (attending to) the mouse engenders a whole-body reaction in Stuart before he is able to derive any sort of meaning from the encounter, yet, epiphany such as this, arriving to the mind through the body, also serves to provide an opportunity 'to recognize and honor nonhuman nature as a world we did not create, a world with its own independent, nonhuman reasons for being as it is', the sort of recognition that is crucially important to environmental historian William Cronon (1995, 87). Murdoch designates intrinsic value to the life of the mouse that does not stem from any human interest. In Murdoch's desire to emphasize that the mouse '*lived there*' (*GA* 480), she reaffirms the significance and intrinsic value of the other-than-self, the more-than-human world, and our utter dependence on such a world.

Murdoch takes a quite different line with the 'reclusive but notorious, dissolute painter', Jesse Baltram (Martin and Rowe 2010, 141), who she sets in culturally mediated nature through his art-making. His art, according to Martin and Rowe, expresses an evil existence as 'an unregulated operation of base desires unmediated by any influences from the outside world' (2010, 141). Alone at Seegard one day, Edward enters his father's tower. And, concerned that he might get '*trapped inside*' (*GA* 194), he nonetheless ventures into the gallery space designed to let in the natural light, to find 'dusty and un-looked-at' exhibits (*GA* 195). Here, he finds strange sculptures of 'entwined pairs, some human, some human and animal', and disturbingly violent yet erotic paintings, including an Ovidian image of 'a youth watching a screaming girl becoming a tree' (*GA* 195). The appalling images of 'big grotesque heads of women, mournful, tearful or vindictive' mingle with depictions of 'battles [..] transmuted into erotic tangles' and 'drowned animals, appalling adolescents, callous or terrified witnesses, deformed people sitting quietly together, stunned by hopelessness and fear, sometimes now watched through doors or windows, by beautiful children, heartless, probably soulless carrying emblems, flags or flowers' (*GA* 195–196). Such exhibits picture depraved art-making, devoid of any form of base morality. These mediated images of the entwined flesh of the world appear to echo Merleau-Ponty's immersive ontology in a manner the philosopher did not intend, but here they are, taken to a bleakly uncompromising, chaotic and amoral conclusion.

In Murdoch's endeavour to create realistic characters, she conceives fully formed and intricate spaces for them to inhabit. Lovibond observes that her novels seem to 'draw a vitality from the evocation of a very precisely visualized material space […] especially—though not only—where characters are pitted directly against natural forces or inanimate objects' (*IMR*5 37). One might add that, despite the fact Murdoch's novels are renowned for being highly populated, it is often in the contrastively solitary moments that characters confront these 'natural forces'—as we have already seen with Effingham Cooper lost in the bog in *The Unicorn* and Tim Reede alone in the mountains of Les Alpilles in *Nuns and Soldiers*. Such moments are suggestive of an interplay or exchange between character and environment, rendering the moment all the more visceral for the reader. Characters collide with natural forces in these powerful moments of interaction.

'Plants Act'

An associated theoretical discourse considers the agency and related implications of vegetal life, such as that which Edward encounters in the Seegard environment: vegetal agency presents a consideration of the natural forces involved in more-than-human interactions. In *The Good Apprentice*, Murdoch cultivates the 'exceedingly flat' landscape of 'waterlogged fields', a windy 'mournful expanse' beset by mud, a track-side ditch emerges into a 'reedy marshy wilderness' and a 'mass of fuzzy darkness' up on a hill suggests woodland (*GA* 104–6). By contrast, Seegard itself is a 'shrine of [...] elegant austerity' (*GA* 112), and the ordered vegetable garden, one senses, has been wrested from these wilder surroundings. The woodland might constitute the 'work of nature not of man' (*GA* 127), but the trees exist under constant threat of instrumentalization by the menacing presence of the mysterious 'tree men' (*GA* 144). Ilona tells Edward,

> they're cutting trees down on the other side, there's a lot of forestry inland. We don't like them very much. They're poisoning all the fretty chervil along the road with weed killer, and they destroyed some orchids. Still they help us sometimes.
> (*GA* 144)

It is too often too easy to leave the vegetal in novels unexamined. Is it just description? Is it a signalling device to indicate a change of scene for the reader? Does it merely colonize the scenic background to human activity? If vegetation is thought about at all, it tends to be assigned instrumental—not intrinsic—value in the minds of character and reader alike (Marder 2013, 4). We are untroubled by plant life in a novel because 'Western culture has always tended to see plants as uninteresting and their existence as unproblematic', argues Salwa (2014, 318). As winter moves into spring at Seegard, the vegetal engenders new life that contrasts with the drab, sterile and joyless descriptions of the hostile London Edward has left behind: 'big indifferent anonymous London' (*GA* 441) had meant to him a dismal, lonely hell.

Murdoch's expressions of vegetal life in *The Good Apprentice* amount to vegetal agency. Prudence Gibson and Baylee Brits argue for what they call '[p]lant thinking', which, they say, 'refers to moving agency away from the human and towards vegetal life', because plants are 'the backbone of all ecosystems', and that 'discounting plant life is a grave ecological and philosophical error' (2018, 16). Contemporary phenomenologist Michael Marder lends metaphysical weight to such 'vegetal exuberance', and Plato (to whom Murdoch's thought is indebted) is expressive on this matter in the *Timaeus* and also informs Marder's thinking (2013, 22).[27] In the *Timaeus*, Plato assigns to trees and plants the status of 'living creature', attributing 'sensation' and 'desire' to them:

> Blending it with other shapes and senses they engendered a substance akin to that of man, so as to form another living creature: such are the cultivated trees and plants and seeds which have been trained by husbandry and are now domesticated amongst us; but formerly the wild kinds only existed, these being older than the cultivated kinds. For everything, in fact, which partakes of life may justly and with perfect truth be termed a living creature. Certainly that creature which we are now describing partakes of the third kind of soul, which is seated, as we affirm, between the midriff and the navel, and which shares not at all in opinion and reasoning and mind but in sensation, pleasant and painful, together with desires. (Plato 1925, 77a–b)

---

[27] Murdoch talks in terms of her agreement with Plato early on in 'On "God" and "Good"' where she aligns her thinking with Plato's starting point of 'love' (OGG 361), and in 'The Sovereignty of Good over Other Concepts' where she explains Plato's use of the sun to explain 'Good' (SGC 375–6); in fact, there are a considerable number of references to Plato in Murdoch's philosophical works that speak to his influence. Broackes assesses Plato's influence on Murdoch (2012, 61–63).

Planting is often perceived as passive landscape for human contemplation or it represents surplus for human benefit. Yet, the 'desires' of plant life do appear plausible immediately one contemplates the compulsion to nourishment or light. In Marder's view, the ontology of plant life carries far greater significance: 'Despite their undeniable embeddedness in the environment, plants embody the kind of detachment human beings dream of in their own transcendent aspiration to the other, Beauty, or divinity' (2013, 12). Observing their 'process of incessant proliferation', Marder believes that 'plants act':

> It would seem that plants act of this desire in the most literal sense, by branching out in all directions: growing in height, spreading horizontally across vast expanses, [...] and imbibing everything from the water, the air, and the soil that surrounds them. (2013, 40)

In essence, when we perceive the world through our senses, we are, in effect, spoken to by landscape, by nature. Vegetal agency is observable in *The Good Apprentice* at crucial points of 'intra-action' (Barad 2007, 141).

As a point of comparison, warranting brief mention at this juncture is the depiction of place in an earlier Murdoch novel, *The Bell* (1958), published almost three decades before *The Good Apprentice*. Both novels portray, in some sense, enclosed orders. The earlier novel is centred around a carefully manicured Capability Brown landscape and lake, and a market garden whose tilling is the subject of debate: machine versus manual labour with the associated considerations of self-imposed austerity and virtuous work. For her purposes, Murdoch presents such notions as romanticized here:

> Toby was a town boy, and everything to do with the countryside had for him a profound, almost spiritual significance. Of sun and wind and hard physical work and human companionship he felt he could never have too much. Given a spade and told to dig up an entire field he would think himself in heaven. (*TB* 43)

And the naked Toby at the water's edge—his 'very pale and slim body was caressed by the sun and shadow as the willow tree under which he stood shifted slightly in the breeze'—completes the 'pastoral vision' (*TB* 75).

Dora Greenfield is reminded of Donatello's David: 'casual, powerful, superbly naked, and charmingly immature' (*TB* 75). *The Good Apprentice*, on the other hand, presents a setting of smallholding and managed forestry, attempting order in the contingent wild, and one can sense Edward's discomfort at being 'unused' to the country (*GA* 124). Within the 'Seegard city state' (*GA* 136), life is regimented and thus reminiscent of Imber Court:

> Edward had become used to the routine of steady ceaseless work punctuated by strictly timed periods of rest, as in a religious order, a monastery, where a good innocent quiet life goes steadily and monotonously on. (*GA* 135)

Outside Seegard, order gives way to an overwhelming sense of the wild from a 'racing substantial river [...] responsible for the flooded water meadow', running 'deeply and swiftly between high steep banks, churning and foaming along with a humming hissing sound', leaving submerged bridges and a 'precarious walkway' in its path—its significance indicated by the richly metamorphotic theme that runs through the novel and inhabits both character and space (*GA* 130).

Edward's own change in circumstances, triggered by causing Mark's death, where Edward 'watched the metamorphosis with wicked triumph' of his friend, is only the beginning of his personal transformation (*GA* 1). Edward is overcome with remorse. Thomas McCaskerville, Edward's uncle and a psychiatrist, 'understands the process of recovery from trauma' (White 2023, 120); Thomas tells his wife Midge, 'It's like a chemical process, Edward has got to change and we have to be, for a time, spectators of that change' (*GA* 41). Thomas believes that Edward will learn to live with the pain. In London, Edward insists that while he feels a sort of change it is a sort of reverse metamorphosis:

> All right, I'm changing, but not in a good way, there is no good way, that's what I've *discovered*. It's not like being—like being a chrysalis—it's the opposite, it's the story run backwards. I used to have coloured wings and fly. Now I am black and I lie on the ground and quiver. Soon the earth will begin to cover me and I shall become cold and be buried and rot. (*GA* 76–77)

Once at Seegard, however, Edward's experience of being alone is one that feels quite different. While Murdoch acknowledges that '[g]oing to Seegard was an ordeal for Edward', she insists that '[o]ne's life can change by a particular drama' (*TCHF* 206). We witness the 'intra-actions' (Barad 2007, 141) between character and landscape when he is out walking. When Edward navigates his solitary way through the flatlands, woodlands and waterlands of Seegard's surroundings, he interacts and reacts to the immersive environment. His body is momentarily almost subsumed into the 'treacherous elastic surface' (*GA* 126) of the flooded landscape as he makes his way towards ancient woodland. These trips into the woods come to represent a freedom for Edward that juxtaposes the strict regime at Seegard, where he is 'after all a prisoner, a prisoner with the kindest, most beautiful, most loving captors' (*GA* 163). He makes illicit trips down to the river and the rushing water overpowering the wooden walkway affords him a thrill akin to 'sexual excitement' (*GA* 163) and he begins to feel 'healthier, stronger, and [he] wonder[s] whether this were a true and proper and *natural* […] recovery' (*GA* 164). When Murdoch highlights these moments, the agential power of Edward's natural surroundings is made evident.

Murdoch's focus on metamorphosis and transformation draws attention to the uncontrolled and overpowering creep of the vegetation in the novel. Early on, during his stay at Seegard, Edward determines to reach the sea and, alongside his desire, we sense urgency. At first, he misses the turning onto a disused railway line that he has seen marked on a local map indicating a route to the coast but, retracing his steps, he eventually finds the way signified by the 'long white gates' that are 'sunk off their hinges and grown over with brambles' (*GA* 231). Although depicted as overwhelmed by vegetation, this turning and the path of the former railway persists in the Seegard landscape—the route for the moment has 'kept its identity' as Edward had hoped and still provides an 'unassimilated way through' (*GA* 228–29). Murdoch's careful lexical choice of 'unassimilated' here suggests that, left alone, the disused railway line would eventually be subsumed by vegetal regrowth and the land returned to a pre-populated state (*GA* 229). As Edward makes his way, 'the work of man' is 'everywhere to be seen in the form of meticulous cultivation' (*GA* 232), implying a reliance on modern machinery, rather than ages-old manual labour, working constantly to push back against the encroaching landscape. But Murdoch describes Edward's disused path as a 'ghost railway line' (*GA* 229), suggesting a less isolated, more populous past and contributing to the sense of the region's present desolation.

One discerns the vegetal gloriously propagating at the edges of a still resolutely defined yet disused and redundant rail route. Edward is overwhelmed in this moment, just as he had been on the solitary woodland walk that occurs at the beginning of his stay at Seegard. Now, the 'sun appeared [..] between scurrying clouds, showing the details of the drooping grasses loaded with silver water, and the velvet texture of the primrose flowers' (*GA* 231). Edward's body enacts an embodied affective response to unalloyed beauty: 'Edward felt as if his heart would burst out of his breast with a great inapprehensible anguish' (*GA* 231). Murdoch depicts Edward's embodied affective response enacted in advance of his mind, before offering full access to his interiority:

> [H]ow can I imagine things about 'recovering' or 'being cured' when what I simply am is *mad*, I have lost my *senses*, I walk along a mad thing, boiling with emotions and pain. Will it always be like this, all of my life, when I am alone and when I see anything beautiful or innocent or good? (*GA* 231–32)

The grasses and the spring-flowering wild woodland primroses provoke at first a visceral, physical reaction in Edward and then an emotional outpouring of grief and self-loathing. Edward has, in effect, been spoken to by his surroundings. This fleeting moment of attention to what is beautiful and good engenders a pure and powerful pain in Murdoch's protagonist, who is yet to imagine any sort of resolution to his despair. Edward's task is to learn to function in a world of his own making after Mark's death.

His difficulty resides firmly in the present, but Murdoch creates a landscape around him that resonates from the long past and into the future; her purpose is to contextualize such suffering into a broader temporal notion of the transitory and insignificant nature of life and human existence: 'just tiny things in someone's glass' (*GA* 172), as Edward suggests to Ilona. On the same walk, at the disused railway station of Smilden Halt, Edward encounters a forlorn cottage that has 'spriggy plants' growing out of the roof beside a 'large deformed yew tree' (*GA* 233). Realizing that the cottage is inhabited, despite appearances to the contrary, he stops to ask for directions and is shocked to find not only Sarah Plowmain answering the door, the young woman for whom Edward had found cause to leave his tripping friend unsupervised, but also Sarah's mother and Mark's sister. Edward reveals the truth of that fateful day's events to his silent, hostile audience in a room 'full of doom and dread and catastrophic forces held in suspense' (*GA* 238). Asked to leave, Edward's relief is palpable as he steps outside 'into the amazing outside air, seeing with astonishment the landscape, just as it had been before, sunlit now, silent, empty, the sun picking out at a distance the soft pale green of a sloping field' (*GA* 238). His relief is short-lived, overwhelmed as he is by a sick guilty pain at the idea that Mark's family are at hand, that this is the cottage where he (Edward) was probably conceived, and his late mother's friend, Sarah's mother, had bought to spite Edward's father.

Late in the novel, Edward returns for a third time to Railway Cottage which is by this time abandoned, and he is inclined to believe he can actually heareeews the vegetal engulfing the cottage building:

> The place was bare, rotting, ruined, soon to be overtaken by weather, by nature, by fungus and green intrusive shoots. As Edward stood and listened he fancied he could hear the soft murmur of this intrusion, the yew trees scraping against the window, the ivy lifting the slates, the insects working deep inside the wood. (*GA* 462)

When Edward leaves the house, the track becomes 'more overgrown with nettles and clumpy sorrel' and Edward can 'feel under his feet the hard stoney surface on which the grass was growing' (*GA* 463). The Railway Cottage represents an instance of the unsettled temporal spectrum of the novel with the focus on its transformation engendered by vegetal agency.[28] Similarly, the fishing harbour once served by the railway has been reclaimed by the sea:

---

[28] This is not the only disused railway line that has been overwhelmed by vegetation in a Murdoch novel. In an earlier novel, *A Fairly Honourable Defeat* (1970), Morgan Browne and her nephew Peter Foster stop the car next to a gulley on the way back from Cambridge and Peter finds an abandoned line. Morgan muses that it 'is a place, a human place, and yet not any more, it has been taken

the village itself he could at first see no trace. Then gazing about he began to discern here and there whole large fragments of stone walls, leaning over at strange angles as in mediaeval pictures of destroyed cities, surrounded by watery pits and overgrown by ivy and wild buddleia. (*GA* 464)

After Jesse mysteriously drowns, towards the end of the novel, Edward returns to Seegard and seeks out his father's gravestone: 'Already the earth had invaded the sides of the stone a little. Later on, if no one tended it, it would become covered with earth, overgrown with grass, lost' (*GA* 514). A gravestone submerged by vegetation certainly points to human insignificance in relation to vaster temporal scales. Yet, presenting at once a separate temporal dimension, Seegard is bathed in soft light and Edward is given the chance to reflect on the wheel of time : 'I have seen the seasons change and the year turn in this place' (*GA* 506). '[W]ild roses [are] profusely in flower', while the cow parsley—Ilona's 'fretty chervil'—(*GA* 506) is over for another year. Murdoch invokes Hopkins's '*Justus quidem tu es Domine*' ['Thou art indeed just, Lord'] for a second time in the novel, this time in recognition of the restorative effects that time in nature has evidently bestowed upon Edward. Hopkins's poem exhorts his reader to take god-given ('O thou lord of life') inspiration from the renewal of life that emerges from the river banks and the nest-building in the trees (*GMH* 67). The celebratory atmosphere of nature's cyclical processes

---

over, lost to us, taken by, yes, by *them*'—who is meant by '*them*' is left enigmatically unclear (*FHD* 176). The narrative describes flowers 'which the scientific farmer had long banished from his fields [that] lingered here in secret, dazing with their variety the drunken bees who crawled laboriously among the stems, buzzing as they walked with sheer exhausted joy' (*FHD* 176). Morgan seems able to conjure names for all the wild flowers 'from far away, out of childhood', and now she looks upon them with wonderment, 'How extraordinary flowers are', followed by the much cited line: 'People from a planet without flowers would think we must be mad with joy the whole time to have such things about us' (*FHD* 177). This piece of narrative is concerned, at once, with the life-giving benefits of rewilding, the negative impacts of pesticides and chemical fertilizers, the beauty of wild flowers, and the importance of being able to name what we seek to protect. This last has become the focus of a noteworthy book, Robert Macfarlane and Jackie Morris (2017) *The Lost Words: A Spell Book*; it was conceived when author and illustrator realized that words such as 'acorn', 'bluebell' and 'kingfisher' had been removed from a widely used children's dictionary, 'because those words were not being used enough to merit inclusion' and yet, as Macfarlane explains, names 'can help us see and they can help us care. We find it hard to love what we cannot give a name to. And what we do not love we will not save'; see https://www.thelostwords.org/lostwordsbook/ (accessed 19 November 2022).

provides a counterpoint to the unstoppable wild vegetal regrowth that even subsumes the evidence of human existence given the chance.

In *The Good Apprentice*, depictions of solitary moments in landscape evoke the affective experience of body immersed or enmeshed in the vegetal. It is significant that Murdoch expresses representations of the natural world in the novel through a form of sensory perception, felt through the body, and she gives full affective expression to the manner in which Edward perceives his environment by means of his senses. At the same time, she acknowledges vegetal agency. Left untended, buildings, railway lines and gravestones become subject to the creep of vegetation, are transformed and overwhelmed. For Barad, the 'intra-action' of character and environment reveals the ethical value (2007, 141). Enmeshment in this way produces an equity that infers a care and respect through the aesthetic appreciation of nature, when we are able to treat 'the elements of nature as subjects' (Salwa 2014, 317). When the natural world is regarded in this way, 'we can grasp for example its emotional values which are objective and are not mere projections of human states of the soul' (Salwa 2014, 317). *The Good Apprentice* can be read as a simple tale of nature cure, where nature cure is an intrinsic element of Murdoch's ethical message.

In this 'otherworldly setting', as Frances White characterizes Seegard, Edward is able to make 'his painful journey from chronic remorse towards lucid remorse' and this places him along the road to recovery (2023, 115). He has been on a quest for forgiveness (which, after some certain rebuttal, he eventually receives from Mark's mother) and on a quest for salvation. But what constitutes salvation in a contemporary post-theistic age? As White asserts, Murdoch has 'divorce[d] salvation from any religious context' in this novel, and yet, the novel asks, 'What, in a secular context, […] can be said to save someone?' And, indeed, 'what, in a secular context, does somebody need saving *from*?' (2023, 126). However one is tempted to try to answer these questions, the time spent at Seegard figures recovery for Edward: 'Recovery from remorse involves both a spiritual need—forgiveness, and a medical need—healing' (White 2023, 115). Edward's immersive experience of woodland and marsh at Seegard contributes to this healing. At the end of *The Good Apprentice*, Edward's extravagant declaration, 'I'm letting go and nature is curing me!' (*GA* 555), is nonetheless tinged with irony:

A picture of ordinary happiness came to him suddenly as a blue sea and a jostle of boats with huge coloured stripy sails. He thought, it's not like what Thomas said about new being and so on, it's more like what he said about the natural ego growing again! (*GA* 555)

It is an irony that Edward, himself, recognizes and that Murdoch highlights about the human psyche. Moments of clarity and humility often prove fleeting.

Murdoch foregrounds her own conception of the embodied mind that combines with agential reality in the vegetal creep of the novel. The vegetal enacts a connection with Edward, serving to emphasize to her reader the transitory nature of life and life's predicaments, which in turn provides Edward with cause to survive and recover. Murdoch demonstrates her deep regard for other sentient beings and organisms on animate Earth and draws our attention to our inescapable relationship to them and our responsibility for them. 'The health of the planet rests upon the health of the individual' (*GA* 173), Mother May is heard to pronounce, but Murdoch seeks to paint a more reciprocal and unifying picture, whereby the health of the individual rests upon the health of the planet, too, her conception of co-existence driven by a love of the contingent beauty of the sacred world.

References

Primary Sources: Novels, Philosophical Writings

Murdoch, Iris. 1997. 'The Existentialist Hero' (EH). In *Iris Murdoch, Existentialists and Mystics: Writings on Philosophy and Literature*, ed. Peter J. Conradi, 108–115. London: Penguin.

———. 1997. 'The Novelist as Metaphysician' (NM). In *Iris Murdoch, Existentialists and Mystics: Writings on Philosophy and Literature*, ed. Peter J. Conradi, 101–107. London: Penguin.

———. 1999. *Sartre: Romantic Rationalist* (SRR) (1953, 1987). London: Vintage

———. 1999. *The Sea, The Sea* (*TSTS*) (1978). London: Vintage

———. 2000. *The Nice and the Good* (*NG*) (1968). London: Vintage

———. 2000. *The Good Apprentice (GA)* (1985). London: Vintage
———. 2001. *A Fairly Honourable Defeat (FHD)* (1970). London: Vintage
———. 2003. *Metaphysics as a Guide to Morals (MGM)*. London: Vintage.

Primary Sources: Journals and Other Writings from the Iris Murdoch Collections

Murdoch, Iris, Journal 3 (J3), 4 June 1945–12 May 1947, KUAS202/1/3
———. 1982. 'Gifford Lectures: 2nd draft', (Giffords) KUAS202/6

Secondary Sources: Books, Chapters, Essays

Bakewell, Sarah. 2016. *At the Existentialist Café: Freedom, Being and Apricot Cocktails*. London: Vintage.
Barad, Karen. 2007. *Meeting the Universe Halfway: Quantum Physics and the Entanglement of Matter and Meaning*. Durham, NC: Duke University Press.
Bate, Jonathan. 2000. *The Song of the Earth*. London: Picador.
Brits, Baylee, and Prudence Gibson. 2018. Introduction: Covert Plants. In *Covert Plants: Vegetal Consciousness and Agency in an Anthropocentric World*, eds. Prudence Gibson and Baylee Brits. Santa Barbara, CA: Brainstorm Books.
Broackes, Justin, ed. 2012. *Iris Murdoch, Philosopher: A Collection of Essays*. Oxford: Oxford University Press.
Broad, C. D. 1937. *The Mind and its Place in Nature* (1925). London: Kegan Paul, Trench, Trübner & Co.
Brown, Charles S., and Ted Toadvine, eds. 2003. *Ecophenomenology: Back to the Earth Itself*. Albany, NY: State University of New York.
Browning, Gary. 2018. Introduction: Interpreting Murdoch—Truth and Love Revisited. In *Murdoch on Truth and Love*, ed. Gary Browning, 1–20. London: Palgrave Macmillan.
Chappell, Sophie Grace. 2022. *Epiphanies: An Ethics of Experience*. Oxford: Oxford University Press.
Clark, Andy. 2008. *Supersizing the Mind: Embodiment, Action and Cognitive Extension*. Oxford: Oxford University Press.
Conradi, Peter J. 2001. *Iris Murdoch: A Life (IMAL)*. London: HarperCollins.
———. 2001. *The Saint and the Artist (S&A)*. London: HarperCollins.
Cooper, David. 2000. Introduction. In Iris Murdoch, *The Good Apprentice* (1985). London: Vintage

Cronon, William. 1995. *The Trouble with Wilderness; or, Getting Back to the Wrong Nature*. In *Uncommon Ground: Toward Reinventing Nature*, ed. William Cronon, 69–90. New York: Norton Publishing.

Denham, A. E. 2012. Psychopathy, Empathy, and Moral Motivation. In *Iris Murdoch, Philosopher*, ed. Justin Broackes, 325–352. Oxford: Oxford University Press.

Dooley, Gillian. 2022. *Listening to Iris Murdoch*, Iris Murdoch Today. London: Palgrave Macmillan.

Fiddes, Paul S. 2022. *Iris Murdoch and the Others: A Writer in Dialogue with Theology*. London: T&T Clark.

Fredriksson, Antony. 2022. *A Phenomenology of Attention and the Unfamiliar: Encounters with the Unknown*. London: Palgrave Macmillan.

Gardner, W. H., ed. 1953. *Gerard Manley Hopkins: A Selection of his Poems and Prose* (*GMH*). London: Penguin.

Goodbody, Axel, and Kate Rigby, eds. 2011. *Ecocritical Theory: New European Approaches*. Charlottesville, VA: University of Virginia.

Gregg, Melissa, and Gregory J. Seigworth, eds. 2010. *The Affect Theory Reader*. Durham NC: Duke University Press.

Hardy, Rob. 2010. Stories, Rituals and Healers in Iris Murdoch's Fiction. In *Iris Murdoch and Morality*, ed. Anne Rowe and Avril Horner, 43–55. Basingstoke: Palgrave Macmillan.

Horner, Avril, and Anne Rowe, eds. 2015. *Living on Paper: Letters from Iris Murdoch 1934–1995* (*LoP*). London: Chatto & Windus.

Iovino, Serenella, and Serpil Oppermann, eds. 2014. *Material Ecocriticism*. Bloomington: Indiana University Press.

James, Henry. 1948. *The Art of Fiction and Other Essays* (1884). New York: Oxford University Press.

James, Simon P. 2018. Martin Heidegger. In *Key Thinkers on the Environment*, eds. Joy A. Palmer and David E. Cooper. London: Routledge.

Kant, Immanuel. 1999. *Critique of Pure Reason*, trans. Paul Guyer and Allen Wood. Cambridge: Cambridge University Press.

Lewis, Charlton T., and Charles Short, eds. 1879. *A Latin Dictionary (OLD)*. Oxford: Oxford University Press.

Lipscomb, Benjamin J. B. 2022. *The Women Are Up to Something*. Oxford: Oxford University Press.

Lovibond, Sabina. 2014. Baggy Monsters Digest the 1980s: The Realism of the Later Iris Murdoch. *Iris Murdoch Review* 5 (*IMR*5)

Mac Cumhaill, Clare, and Rachael Wiseman. 2022. *Metaphysical Animals: How Four Women Brought Philosophy Back to Life (MA)*. London: Chatto & Windus.

Macfarlane, Robert. 2007. *The Wild Places*. London: Granta.

Macfarlane, Robert, and Jackie Morris. 2017. *The Lost Words: A Spell Book*. London: Hamish Hamilton.

Marder, Michael. 2013. *Plant-thinking: A Philosophy of Vegetal Life*. New York: Columbia University Press.

Martin, Priscilla, and Anne Rowe. 2010. *Iris Murdoch: A Literary Life*. Basingstoke: Palgrave Macmillan.

Merleau-Ponty, Maurice. 1964. *Signs*, trans. Richard McCleary. Evanston, IL: Northwestern University Press, IML 1069, Iris Murdoch Collections.

———. 1968. *The Visible and the Invisible: followed by Working Notes*, trans. Alphonso Lingis. Evanston, IL: Northwestern University Press, IML 541, Iris Murdoch Collections

———. 1973. *Adventures of the Dialectic*, trans. Joseph Bien. Evanston, IL: Northwestern University Press

———. 2012. *Phenomenology of Perception* (1945), trans. by Donald A. Landes. Abingdon: Routledge.

Oulton, Lucy. 2014. '*Being in the presence of beauty*': *Art, Affect and the Embodied Mind in Novels by E. M. Forster, Iris Murdoch and Zadie Smith*. Unpublished masters dissertation, Kingston University.

———. 2022. Nature and the Environment. In *The Murdochian Mind*, eds. Silvia Caprioglio Panizza and Mark Hopwood, 453–467. London: Routledge.

Plato. 1925. *Timaeus*, Plato in Twelve Volumes, Vol. 9, trans. W.R.M. Lamb. London: William Heinemann

Purton, Valerie. 2007. *An Iris Murdoch Chronology*. Basingstoke: Palgrave Macmillan.

Rowe, Anne. 2019. *Iris Murdoch*, Writers and Their Work. Liverpool: Liverpool University Press.

Salwa, Mateusz. 2014. The Garden—Between Art and Ecology. *Proceedings of the European Society for Aesthetics* 6: 316–327.

Sartre, Jean-Paul. 1963. *Nausea*, trans. Robert Baldick. London: Penguin.

Shepherd, Nan. 2011. *The Living Mountain* (1977). Canongate: Edinburgh.

Stevenson, Angus, ed. 2010. *Oxford Dictionary of English*, 3rd ed. (*OED*). Oxford: Oxford University Press.

Waugh, Patricia. 2012. Iris Murdoch and the Two Cultures. In *Iris Murdoch: Texts and Contexts*, eds. Anne Rowe and Avril Horner, 33–58. Basingstoke: Palgrave Macmillan.

Weik von Mossner, Alexa. 2017. *Affective Ecologies: Empathy, Emotion, and the Environmental Narrative*. Columbus, OH: Ohio State University Press.

Westling, Louise. 2014. *The Logos of the Living World, Merleau-Ponty, Animals, and Language*. New York: Fordham.

White, Frances. 2023. *Iris Murdoch and Remorse: Past Forgiving?* London: Palgrave Macmillan.

6

A Vision of the World as Sacred: Further Thoughts on Murdoch's Ecological Consciousness

'TIME PASSES AND WE MUST THINK ABOUT THE ECOLOGICAL DISASTER WHICH IS BY NOW INCREASINGLY VISIBLE'

In *Under the Net*, Jake Donaghue likens starting a new novel to 'opening a door on a misty landscape'; he says, 'you can still see very little but you can smell the earth and feel the wind blowing' (*UN* 277–8). Murdoch's analogy is undemanding: she ascribes the sheer adventure of uncharted lands to this new work of fiction. It represents the straightforward opportunity to stop, to look and to feel; it conveys the exhilaration that awaits those ready to immerse themselves and their senses in a metaphorical, or a real, landscape. Such is the nature of Murdoch's own immersive experience of the living world. She found her garden and the wild places where she walked and swam spiritually uplifting, and she animated her philosophy of attending to the Good by allusion to the natural world, most notably through her illustration of the hovering kestrel. Peter J. Conradi's obituary of Murdoch for the *Guardian* confirms her worldview: 'God and the after-life were essentially anti-religious bribes to her', he wrote; 'her vision of the world as sacred looks forward to ecology and the Green movement' (Conradi 1999). In the absence of a unified belief in a transcendent God and turning instead to a conception of the Good, Murdoch sought immanent presence in the material world around her.

Murdoch's oeuvre expresses a reverence for vibrant Earth, our physical and psychological dependence on it and, simultaneously, our human frailty in relation to it. However, in the twentieth century, moral consideration for Murdoch extended to a disquiet about a world imperilled by humankind's inaction or wrong action—a concern that takes on yet greater proportions today. Her novels demonstrably highlight such themes when her representations of the physical environment bring her affinity with, and deep concern about, the future of Earth progressively into the light. While Murdoch's human characters, their behaviours and their interactions are unarguably positioned front and centre of her fiction, there are significant scenes in many of the novels where her characters are portrayed as fully immersed in the materiality of the living world, and these function to articulate Murdoch's environmental imagination. These traits in her fiction have their beginnings in her poetry. Here, she delineates her belief in an essential symbiosis of a healthy planet with all life, inciting in readers a care and concern for the physical world. In her poems she builds a sense of belonging to and reliance on the landscape she pictures and, with them, she seeks to convey the integral and immersive involvement of our environment and ourselves. In the novels, she pictures solitary moments of epiphany, the individual in discursive communion and exchange with nature. *Iris Murdoch's Wild Imagination* has sought to establish an entry point for further discussions of these depictions of interchange with the natural world; there remains much to consider in this area of Murdoch studies and I outline here three aspects of potential or ongoing interest that call for further research in relation to her work.

Murdoch, Ancient Spiritual Traditions and Deep Ecology

In the latter half of her life, Murdoch became increasingly interested in ancient spiritual, philosophical and religious traditions; she was particularly interested in Buddhism, but books on Taoism and Hinduism populate her libraries.[1] For a time, she was setting out to examine such spiritual practices on her quest for a neo-theology; more specifically, she sought to engage with non-theistic religions. For many of these traditions (as it is with many indigenous practices) Earth is seen as sacred, the value of the natural world transcending human schemes; some of these ideas foreground the inherent rights of all living things to a standing in the world; some go so far as to regard all food as sacred (Barnhill and Gottlieb 2001, 9). Such values resonate with those who champion the ecocentrism of the deep ecology movement.

Deep ecology as a concept represents a multivalent set of values as I have mentioned and, for these purposes here, is considered again in broad-brush terms. The movement's inception in 1973 is attributed to Norwegian philosopher and mountaineer Arne Naess; deep ecologists reject capitalism in favour of a model that is able to recognize the 'inherent value of all living beings and the use of this view in shaping environmental policies' (Drengson 2012). For context, a shallow approach to ecology, arguably one more likely to win hearts and minds, addresses sustainable fuel sources, reuse and recycling but continues to prop up existing societal and governmental systems, but this falls short for deep ecologists; deep ecology's target is major systemic change. Deep ecologists argue that the current drive towards sustainability may have the prospects for human populations in their sights, but fails to take account of broader ongoing devastating effects on the planet such as deforestation, habitat loss and other related issues threatening many species. Deep ecologists argue that these issues cannot be tackled in isolation.

Murdoch's objections to the creep of technology and, more generally, to the pervading scientism of the modern age have drawn one critic to

---

[1] Murdoch's libraries in Iris Murdoch Collections include a number of works; some are (heavily) annotated and these are selected examples: Ananda Kentish Coomaraswamy (1971) *Hinduism and Buddhism*. Westport CT, Greenwood Press, annotated, IML 308; Eugen Hemgel (1962) *The Method of Zen*. London, Routledge and Kegan Paul, IML 417; Katsuki Sekida (1975) *Zen Training: Methods and Philosophy*. London, Weatherill, annotated, IML 50; Alan Watts (1957) *The Way of Zen*. London, Thames and Hudson, annotated, IML 52; Robert Charles Zaehner (1966) *Hinduism*. Oxford, Oxford University Press, IML 324; Daisetz Teitaro Suzuki (1963) *Outlines of Mahayana Buddhism*. London, Schocken Books, IML 295; Fritjof Capra (1976) *The Tao of Physics*. London, Fontana, heavily annotated, IML 54; and two editions: Laozi (1945) *Tao Te Ching*. London, Buddhist Society, IML 796, and Laozi (1972) *Tao Te Ching*. London, Allen and Unwin, IML 410.

align Murdoch's ecological concerns with those of physicist Fritjof Capra, author of *The Tao of Physics* (IML 54). According to Zeynep Yilmaz Kurt (2020), Murdoch's anxiety about technology's threatening pace—'[t]echnological changes which used to be slow and invisible are now fast and perceptible' (EM 221)—bears a certain equivalence to Capra's position that 'nuclear weapons [...] threaten to wipe out all life on the planet, toxic substances [...] contaminate the environment on a large scale, new and unknown microorganisms [await] release into the environment without knowledge of the consequences' (1976, 20). What appears to have been overlooked here, however, is Murdoch's incisive dismissal of Capra's 'popular metaphysic' (*MGM* 198) in *Metaphysics as a Guide to Morals*.[2]

Murdoch's rejection of Capra's *Tao* is located in the same chapter of this work where she airs her frustrations with Derrida. She sets what she perceives as the 'amoral determinism' of the Derridean model—something 'we *should*, [...], in our great technological era and on our smaller more vulnerable planet, be *afraid of*—in juxtaposition with what she regards as popular science in the guise of a lesser-understood 'kind of instinctual debased Taoism' (*MGM* 198). In Murdoch's heavily annotated copy of Capra's bestselling work she has underlined this claim: 'Quantum theory forces us to see the universe [...] as a complicated web of relations between the various parts of a unified whole', which, he maintains, aligns with the long-held views of 'Eastern mystics' (IML 54, 142). Capra asserts that, as 'the rational mind is silenced, the intuitive mode produces an extraordinary awareness; the environment is experienced in a direct way without the filter of conceptual thinking' (IML 54, 40). As I explain in *The Murdochian Mind*, 'even as she acknowledges the part aesthetic experience has to play', Capra's claim simply 'represents the failure of language' as far as Murdoch is concerned (Oulton 2022, 461). One of Murdoch's annotations in blue biro at the bottom of the page in Capra's book asks: 'Apprec[iation] of art, at best, is non-verbal, even in lit[erature]?' (IML 54, 40) Murdoch is able to identify with Capra's sense of the failure of language to convey reality, but considers his acceptance here as capitulation when he declares that the 'experience of oneness with

---

[2] I have discussed Murdoch's assessment of Fritjof Capra's 'popular metaphysic' before (Oulton 2022, 461–62).

the surrounding environment is the main characteristic of this [Taoist] meditative state. It is a state of consciousness where every form of fragmentation has ceased, fading away into undifferentiated unity' (Capra IML 54, 40). In her chapter on 'Derrida and Structuralism' (*MGM* 185–216), Murdoch airs her two frustrations at once. Capra's position, she says, manifests itself as

> a kind of instinctual debased Taoism, arising in a period of exceptional scientific and technological progress and popular scientific knowledge, a relaxed acceptance as ultimate of a deep impersonal world-rhythm which overcomes the awkward dichotomies between good and evil and one individual and another. A sort of neo-Taoism is also part of a popular metaphysic of our time. (*MGM* 198)

Simply expressed, while one philosopher seeks to tie everything up in language, another appears to promote an ontology that might attempt to get by without any language at all. The Tao, as presented by Capra, suggests a world held in natural balance that continuously resets itself, a harmony which forms the basis of the Gaia hypothesis, but it appears, for Murdoch, to constitute little more than wish-fulfilment. While she respectfully acknowledges the vast religious, philosophical and social reach of Taoism, she takes issue with its appropriation and the interpretation of its concepts in 'western books about oriental religion' such as Capra's (*MGM* 199). In short, she believes philosophy cannot settle for 'neo-psychologism' at the expense of ethical discussion (*MGM* 216). Capra went on to publish as a deep ecologist two decades later, but there is no evidence that Murdoch read his later work.[3]

In Murdoch's twenty-first novel, *The Philosopher's Pupil*, Father Bernard is a disillusioned priest struggling with his faith, who seeks to find meaning in his various conversations with enchanter-figure John Robert Rozanov, but unsuccessfully.[4] At the very end of the novel, the priest has penned a faintly comical letter to Murdoch's strangely enigmatic narrator, N, that appears to gush with born-again enthusiasm:

---

[3] For example, Fritjof Capra (1983) *The Turning Point: Science, Society and the Rising Culture*. London, HarperCollins; and Fritjof Capra (1996) *The Web of Life*. London, HarperCollins.
[4] There is an earlier version of this short section on *The Philosopher's Pupil* (Oulton 2022, 462).

> Nothing else but *true religion* can save mankind from a lightless and irredeemable materialism, from a technocratic nightmare where determinism *becomes true* for all except an *unimaginably depraved* few, who are themselves the mystified slaves of a conspiracy of machines. (*PP* 552)

However, Father Bernard's new religion constitutes the '*absolute denial of God*' (*PP* 552). He has taken up a solitary existence by the sea, reliant on villagers for subsistence. He says that everything he sees around him is what is real: 'Only perceive purely and the spiritual and the material world vibrate as one [...]. There is no beyond, there is only here, the infinitely small, infinitely great and utterly demanding present' (*PP* 553). When N reads a section of a letter from Father Bernard aloud, the sensitive Gabriel McCaffrey wipes a tear but husband Brian, we are told, just thinks the priest has gone 'batty' (*PP* 554), diametric reactions that speak, one could surmise, to Murdoch's own equivocation. She is in sympathy with her priest's transformation, but recognizes the work there is to do; humankind has an ethical responsibility towards the natural world. In the closing pages of *The Philosopher's Pupil*, it seems as if Father Bernard has undergone spiritual rebirth as a deep ecologist. While Fritjof Capra's approach to Taoism may have had little appeal for Murdoch, ancient spiritual traditions were themselves of significant interest for Murdoch, and it would be interesting to interrogate, more comprehensively, the degree to which this interest intersects with her environmental imagination.

Murdoch and Environmental Philosophy

In Murdoch's review of ecologist and friend Brian Medlin's book *Human Nature, Human Survival*, published in *The Age* on 27 February 1993, she grimly accepts that the planet is in peril and expresses her belief that we are set on a course of ecological destruction. She apprehends the need for humankind to act, and urgently. Her own bleak conclusion conveys a palpable anxiety:

> What will our poor ailing planet be like in the next century? [...] We have to change ourselves if we are to change the world. <u>Time passes</u> and we must <u>think</u> about the ecological disaster which is by now increasingly visible. (Dooley and Nerlich 2014, 208)

Medlin, Professor of Philosophy at Flinders University, an activist and poet, published his 'passionate monograph' (2014, 199) in 1992, in which he emphasizes the failings of capitalist ideology and the distinct connection to be made between ecological devastation of the planet and the threat to human survival. In his book Medlin (1992, 19) quotes Bradley Pearson's passionate and mocking exposition on the harsh inequalities of human suffering from *The Black Prince*—Medlin's purpose, to admonish smug bourgeois elites:

> This is the planet where cancer reigns, where people regularly and automatically and almost without comment die like flies from floods and famine and disease, where people fight each other with hideous weapons to whose effects even nightmares cannot do justice, where men terrify and torture each other and spend whole lifetimes telling lies out of fear. (*BP* 348–49)

Bradley adds, soberingly: 'This is where we live' (*BP* 349). Murdoch was not interested in Medlin for his Marxist politics. As she explained to him, 'I used to be Marxist (years in the C[ommunist] P[arty]) but no more. I think Marxism is either something generally obvious or else (if it covers the <u>whole</u> horizon) very misleading' (Dooley and Nerlich 2014, 2–3). Communist ideology had lost its appeal for Murdoch decades before in the 1940s. In fact, during the course of her adult life, her politics migrated to such an extent that by the 1980s Murdoch had become a supporter of Margaret Thatcher's Conservative Party. 'Iris's own perception', writes Conradi, 'was not that she had moved to the right, but that the Labour Party had been taken over by left-wing extremism' (*IMAL* 572). Murdoch's political allegiances certainly went through an extraordinary transformation over her lifetime.

She was a loyal friend to Medlin, nonetheless, and in this connection expresses misgivings about reviewing his work: 'there is a difficulty, which is that I disagree with some of your main tenets—the root of which is your sort of Marxism-Leninism and your anti-bourgeois arguments' and instead of a review, she initially proposed a 'counter piece' (Dooley and Nerlich 2014, 177). Where the two philosophers did align was on ecology: 'The other deep stream of your thought, as I see it, is your excellent ecological argument' (Dooley and Nerlich 2014, 177). Murdoch's eventual and frank review also acknowledges this, and testifies to her understanding of the urgency and the enormity of the task that we continue to face today, to secure a liveable and sustainable future for all life.

Murdoch had learnt something of Australia's ecology from Medlin in correspondence with him over the years since first meeting him through Bayley at New College, Oxford, in about 1961, according to Dooley (2014, ix). Occasionally, finding himself in receipt of remarks from Murdoch such as 'I think of your country as mainly <u>untouched wilderness</u>', the ecologist would counter: 'In two hundred years, we have devastated this continent. The full extent of the damage isn't always obvious to the untutored eye [...]. At least two-thirds of our arable pastoral lands are now considered to be seriously degraded' (Dooley and Nerlich 2014, 117, 119)[5]. Many letters travelled back and forth between the pair from 1976 to 1995 until, with the onset of Alzheimer's disease in the latter half of the 1990s, correspondence became impossible for Murdoch: her final letters to Medlin are 'heartbreakingly simple', Dooley observes (2014, xii).[6]

In her review of Medlin's book, Murdoch connects social ills with environmental destruction: 'We have to care on a large scale, there is so much to do, we have to see <u>a bad</u> society, we have to picture a good society' and, she continues, '[p]overty must be fought with, and this too depends upon the fight to preserve the whole planet' (Dooley and Nerlich 2014, 206, 207). Fundamental to all of this is love. 'Love is passion', Murdoch says, 'love is intelligence' (2014, 207). As far as she is concerned, love engenders attention and needs to be placed at the heart of the fight. Murdoch's interest in Medlin and his work helps to inform ideas about the extent to

---

[5] The correspondence between Murdoch and Medlin is published in this volume; see also Dooley's account of Murdoch's difficulties with producing her review for the *Age* (Dooley and Nerlich 2014, xii–xiii).

[6] Dooley recounts how, as part of her research for *Never Mind about the Bourgeoisie*, she went to meet John and Audi Bayley at their home in Charlbury Road, Oxford, to learn more about the context and nature of the friendship between the two philosophers, Medlin and Murdoch (Dooley, 2021).

which Murdoch's philosophy can be brought to bear on moving hearts and minds in relation to the environmental crisis, and there is clearly more work to do in this area.

Murdoch and Non-Human Animals

Murdoch's representations of non-human animals in the novels continue to be a source of fascination to critics.[7] I have discussed the mouse that runs across the rail tracks that captures Stuart's attention in *The Good Apprentice*. Shortly before this scene, a robin flies in through the window during a desperate argument between Thomas and Harry about Thomas's wife Midge with whom Harry has been having an affair; the robin 'began rapidly circling the ceiling, occasionally thumping its frail small body against the walls' (*GA* 459). When the robin falls to the ground and for a moment is ominously still, the two men are stirred to action to rescue the bird, and the situation is diffused. A third notable intervention happens, earlier in the narrative, when Brenda Wilsden (Brownie) has asked Edward to meet her by the river to talk about her brother Mark's death. Beset by gloom as Edward recalls her mother Jennifer Wilsden's hate-filled letters, he worries that Brownie will not appear after all. Suddenly, there is a 'small explosion just above the water, a blue flash' (*GA* 243): a kingfisher briefly interrupts his 'black sick faint feeling' (*GA* 244), diverts his attention and provides a welcome conversation piece when Brownie does appear. Edward finds himself able to tell Brownie exactly what happened before her brother died. Many of Murdoch's novels are resplendent with the comings and goings of such animal life; they are far from incidental and warrant further study. Indeed, the apparently diversionary powers of creatures such as the robin and the kingfisher call to mind Antony Fredriksson's work on intersubjectivity, when they seem to carry the sort of attentive significance that Fredriksson attributes in his own example of his dog drawing his attention to a squirrel (2022, 111–12), noted earlier.

---

[7] While researching sound for *Listening to Iris Murdoch*, Dooley collected all the incidences of birdsong and animal cries in Murdoch's novels; it is to be hoped that this research will feature in her future work.

For Mathilde La Cassagnère, dogs represent Murdoch's 'totem ethical animal' (2020, §8), and she reminds us that Murdoch once said that 'dogs are often figures of virtue' (*TCHF* 155). La Cassagnère proposes a 'zoopoetic approach' to examining animal alterity in Murdoch's novels; she asserts Murdoch seeks to address this through 'zoofocalization', when she passes everything 'through the animal's perceptions' (2020, §7). La Cassagnère discusses Anax from *The Green Knight* and Zed from *The Philosopher's Pupil*. As she points out, Conradi has attributed an 'animal intelligence' to Murdoch, which he defines as an 'ability to encounter the sensuousness of the activity of thinking' (*S&A* 8). La Cassagnère also attributes many of the same qualities of the Murdochian dog to other 'beneficent beings' (*TSTS* 476) that appear in some of the novels: seals, or 'sea-dogs' as La Cassagnère prefers to call them (2020, §7). It is certainly tempting to regard Murdoch's 'cynomorphism' as a cosy, and indeed amusing, attempt at yet another type of character (La Cassagnère 2020, §7). However, an alignment of domestic with wild animals ought to invite interrogation, when wild animals exist independently of, or indeed despite, humankind. There is clear prescience in Murdoch's attempts at another point of view when one considers contemporary efforts to decentre the human.

Murdoch's perceived role in contemporary debates around animal ethics invites further research. Tony Milligan looks at the intersection of ethics and otherness and in his recent essay for *The Murdochian Mind*, writing on 'loving attention to animals', he refers to what he calls the 'mixed intentionality of emotions' (2022, 477). He suggests that unselfing demands an implicit sense of self to begin with and that Murdoch's method, of using Platonic imagery to direct attention outwards, does not appear to fully confront this in her philosophical writing: 'the sense of self in our love for others. A sense which must be there precisely because it is *our* love rather than anyone else's' (Milligan 2022, 477). Milligan avers that 'all love, for dogs and kestrels as much as other humans, is about both reaching outwards, or looking away from the self, and also looking towards self, but in a less obsessive way' (2022, 477). He senses that the novels are more successful at making this point.

Elisa Aaltola notes that 'Murdoch has defended emotions as the foundation of moral agency, as she has posited that emotions colour the world with normative, moral hues', which in Aaltola's view have a bearing on the contemporary 'affective turn' (2022, 67). She examines the work of contemporary affect theorists and neuropsychologists that suggests that 'emotions also help to conceptualise moral matters' (Aaltola 2022, 68). She argues that 'reflective empathy' (the mode of empathy that helps us evaluate 'why and with whom we empathize') enhances 'our ability to evaluate and cultivate those emotion concepts relevant to animal ethics' and she, too, remarks on Murdoch's warning of the 'small world' created by our human tendency to solipsism: 'Egoistic anxiety veils the world' (*MGM* 175). Aaltola avers that 'concentrating principally on one's own immediate benefits all too easily leads to defining others via them: thus, instead of witnessing a forest or a pig, we notice potential paper pulp or bacon'; she concludes that how we value and treat other animals depends partly on how we think about them emotionally (2022, 72, 70, 77). Aaltola goes on to reflect on Murdochian attention and its contribution to reflective empathy, and how we might consider animals more holistically in this context. The contemporary interest of philosophers and literary critics in the changing attitudes towards animals, the perspectives of animals, and the views of animal ethicists working on Murdochian philosophy calls for further investigation of the novels in relation to Murdoch's depictions of non-human animals.

Murdoch's Wild Imagination

Murdoch intuits nature, and both visual and literary art, as foci of human apprehension. However, she frequently delineates nature as the 'more sensuous experience' when, for instance, 'we contemplate formless endless works of nature, waterfalls, mountains, starry heavens'. Such efforts are often rewarded with a sense of awe: 'In nature we find both sublimity and beauty. We are thrilled by its vastness and its sheer chaotic confusion' (*MGM* 311). Murdoch remarks on our tendency to seek consolation in alternative experiences which she attributes to the absence of religion: '"religious experience" now tends to mean intense personal impressions rather than supernatural visions. Also mentioned in such arguments are interchanges with nature, profound feelings in solitary places, when listening to music, and so on' (*MGM* 341). In her philosophy, Murdoch interrogates whether in our demythologized age 'profound feelings in solitary places' can substitute for religion and represent 'moral experiences'; in her fiction, as this book demonstrates, we glimpse such 'interchanges with nature' (*MGM* 341) in which she portrays the whole-body experience of a (usually solitary, often brief) moment of epiphany. These moments are profoundly important to Murdoch.

Iris Murdoch's Wild Imagination invites readers to consider the aesthetic and ethical dimensions of ecological thought when Iris Murdoch, novelist, poet, philosopher and public intellectual, presents her fully immersive approach to the world in both her published and unpublished writings. This book is the first of its kind to trace Murdoch's deep love of the natural world, to identify in her writings an ecological consciousness, and to examine ways in which ecological themes, present from the start of her literary life, pervade her personal journals, her poetry, her philosophy and her fiction. It is also the first of its kind to draw connections between the themes that preoccupied the young poet and those that emerge later in the novels. This study enhances critical awareness in the field of Murdoch studies by presenting innovative readings of a few of her published, and previously unpublished, poems, and by suggesting new interpretive potential for some of the novels in foregrounding her depictions of landscape. At the same time, Murdoch is introduced to a new audience in the already sizeable and diverse field of ecocriticism. The intention is for this work to make its own contribution in highlighting the significance, to ecocriticism, of the uniquely Murdochian approach to fictional realism which takes account of fully animistic landscapes and pictures human immersive, affective interaction therein—specifically, in solitary (albeit often momentary) epiphany. The study also signals how Murdoch's philosophical approaches might contribute to environmental discussions: her ethics of attention centred on love is fundamental here. Further, if one considers that the sense of human primacy over Earth constitutes pure fantasy on a planet of finite resources, then Murdoch's ethics of attention may usefully be set to work on the matter. This book exemplifies what White foresees in her monograph, that the 'dynamic interaction between the novels and readers' lives is Murdoch's distinctive literary-moral achievement which ensures that future reading generations will discover her timeless relevance to their own ethical dilemmas' (White 2023, 203).

Rowe, too, often remarks on the shape-shifting quality of the novels (*Podcast* 6 February 2020) that allows them to speak to a remarkable range of contemporary issues: ecology is no exception. Murdoch's personal writings have contributed significantly here, by drawing attention to and, indeed, confirming the environmental imagination of the novelist that permeates the poetry, the philosophy and the fiction.

The acquisition and assimilation of these personal journals, poetry notebooks and letters have enriched our critical understanding and produced fresh perspectives on Iris Murdoch, the person, and her oeuvre. Her ecological interest has come into the light in this way, inspiring a rereading of her novels and philosophy to trace this preoccupation. Her concern for the environment since her early membership of Men of the Trees, the significance of place and situatedness for Murdoch, her belief in the restorative effects of the natural world, her profound sense of the cathartic, even redemptive, powers of rivers and wild, often dangerous, oceans—indeed, how such places and their representations in literature can tell us so much about ourselves—is tempered by her growing concern about the devastation at the hands of humankind. All of these aspects of Murdoch's environmental imagination are to be found in the various records relating to her life.

References

Primary Sources: Novels, Philosophy, Other Writings

Murdoch, Iris. 1997. 'Existentialists and Mystics' (EM). In *Iris Murdoch, Existentialists and Mystics: Writings on Philosophy and Literature*, ed. Peter J. Conradi, 221–234. London: Penguin.

———. 2000. *The Philosopher's Pupil* (*PP*) (1983). London: Vintage.

———. 2002. *Under the Net* (*UN*) (1954). London: Vintage.

———. 2003. *Metaphysics as a Guide to Morals (MGM)*. London: Vintage.

———. 2013. *The Black Prince* (*BP*) (1973). London: Vintage.

———. 2014. Review of *Human Nature, Human Survival* by Brian Medlin. In *Never Mind about the Bourgeoisie: The Correspondence between Iris Murdoch and Brian Medlin 1976–1995*, ed. Gillian Dooley and Graham Nerlich, 199–208. Newcastle: Cambridge Scholars Publishing.

Secondary Sources: Blogposts, Books, Chapters, Journal Essays, Newspapers, Podcasts

Aaltola, Elisa. 2022. Affective Animal Ethics: Reflective Empathy, Attention and Knowledge *Sub Specie Aeternitatis*. In *Human/Animal Relationships in Transformation: Scientific, Moral and Legal Perspectives*, eds. Augusto Vitale and Simone Pollo. London: Palgrave Macmillan.

Barnhill, David Landis, and Roger S. Gottlieb, eds. 2001. *Deep Ecology and World Religion: New Essays on Sacred Ground*. Albany, NY: State University of New York.

Capra, Fritjof. 1976. *The Tao of Physics*. London: Fontana, IML 54.

Conradi, Peter J. 1999. A Witness to Good and Evil. *Guardian*, 9 February 1999. https://www.theguardian.com/news/1999/feb/09/guardianobituaries-peterconradi. Accessed 21 July 2021.

———. 2001. *Iris Murdoch: A Life (IMAL)*. London: HarperCollins.

———. 2001. *The Saint and the Artist (S&A)*. London: HarperCollins.

Dooley, Gillian. 2021. Visiting Charlbury Road. *Iris Murdoch Society Blogpost*, 3 May 2021. https://irismurdochsociety.org.uk/2021/05/03/visiting-charlbury-road/. Accessed 19 April 2023.

Dooley, Gillian, and Graham Nerlich, eds. 2014. *Never Mind about the Bourgeoisie: The Correspondence between Iris Murdoch and Brian Medlin 1976–1995*. Newcastle: Cambridge Scholars Publishing.

Drengson, Alan. 2012. Some Thought on the Deep Ecology Movement. *Foundation for Deep Ecology*. http://www.deepecology.org/deepecology.htm. Accessed 4 July 2020.

Fredriksson, Antony. 2022. *A Phenomenology of Attention and the Unfamiliar: Encounters with the Unknown*. London: Palgrave Macmillan.

La Cassagnère, Mathilde. 2020. On Dogs and Good: Iris Murdoch's Animal Imagination. *Études britanniques contemporaines* 59. https://doi.org/10.4000/ebc.10137.

Leeson, Miles, Lucy Oulton, Anne Rowe and Frances White. 2020. Under the Net. *Iris Murdoch Podcast*, 6 February 2020.

Medlin, Brian. 1992. *Human Nature, Human Survival*. Adelaide: Board of Research, Flinders University.

Milligan, Tony. 2022. Loving Attention to Animals. In *The Murdochian Mind*, eds. Silvia Caprioglio Panizza and Mark Hopwood, 468–478. London: Routledge.

Oulton, Lucy. 2022. Nature and the Environment. In *The Murdochian Mind*, eds. Silvia Caprioglio Panizza and Mark Hopwood, 453–467. London: Routledge.

Yilmaz Kurt, Zeynep. 2020. Deep Ecology and Representation of the Non-Human in Iris Murdoch's Late Fiction. *Interactions* 29: 1–2. https://www.questia.com/read/1G1-618575651/deep-ecology-and-representation-of-the-non-human. Accessed 10 June 2020.

White, Frances. 2023. *Iris Murdoch and Remorse: Past Forgiving?* London: Palgrave Macmillan.

Index[1]

A

Aaltola, Elisa, 234, 235
Abram, David, 21, 23, 128, 136, 137
  See also More-than-human
Affect theory
  affective engagement, 184
  affective turn, 190, 235
  affect theorists, 23, 24, 34, 235
The Age, 230, 232n5
Amphibians, 159
  toads, 86, 159
Anders, William, 2
Animals
  animal ethicists, 235
  animal ethics, 132, 234, 235
  See also Species

Animism
  animist affinities, 114
  animist enchantment, 115
  animistic landscapes, 34, 88, 97, 185, 236
  animistic sensibility, 33, 78, 96, 97, 128, 129, 140, 175n12
Anscombe, G. E. M. (Elizabeth), 186, 186n5, 188
Anthropocene, 25, 25n17, 26, 163
Anthropocentrism, 21, 22, 130, 161
Anti-scientism, 11
  See also Scientism
Aquatic birds, 158
Aristotle, 8, 162
Arthropods, 159
Artificial intelligence, 10

---

[1] Note: Page numbers followed by 'n' refer to notes.

© The Author(s), under exclusive license to Springer Nature Switzerland AG 2025
L. Oulton, *Iris Murdoch's Wild Imagination*, Iris Murdoch Today,
https://doi.org/10.1007/978-3-031-87833-6

241

Index

Association for the Study of Literature and the Environment (ASLE), 16n10
ASLE-UKI, 16n10
Attention, ethics of, 132, 199n21, 236
attentiveness, 57, 190
Attfield, Robin, 73–75
Auden, W. H., 43n1, 51, 52, 56
  The Enchafèd Flood, 51
  'In Memory of W. B. Yeats,' 56
Austen, Jane, 104
Avebury, 129n12

B

Badminton School, 51, 52
  Badminton School magazine, 45n3, 46, 51, 52n7, 129n12
Barad, Karen, 184, 184n1, 185, 197, 203, 212, 214, 218
  intra-action, 184, 212, 214, 218
Bate, Jonathan, 19, 20, 20n12, 22, 47, 49, 50, 56, 195, 195n16
Bavidge, Jenny, 5, 56
Bayley, Audi, 3n4, 232n6
Bayley, John, 3, 53, 75, 98, 98n4, 126, 133, 138n20, 140, 149, 155, 156, 162, 175n12, 176, 176n17, 232
BBC, 75, 186, 191, 191n11
Beauvoir, Simone de, 190–192
Belonging, 33, 44, 61, 96, 98n4, 99, 110, 128, 157, 164, 226
Bennett, Jane, 23, 24, 135, 137, 140
Betjeman, John, 51, 52
Bigsby, Christopher, 43, 87, 192
Biles, Jack, 30

Birds
  choughs, 158
  cormorants, 158
  curlews, 158
  gannets, 158
  guillemots, 158
  gulls, 158
  kestrels, 15, 63, 73–75, 79–81, 131, 184, 225, 234
  kingfishers, 217n28, 233
  nightingales, 83–86
  oyster-catchers, 158
  robins, 233
  shags, 158
  windhovers, 73, 75, 81
Bladow, Kyle, 23
Blake, William, 6, 45, 46, 60, 65, 65n15, 165
  Book of Thel, 65
  'The Fly,' 165
  'The Sick Rose,' 60
  Songs of Innocence and Experience, 165
  'The Tyger,' 45
Blue ecology, 34, 150
Bolton, Lucy, 95
Booker prize, the, 54
Boston University Journal, 53
Bove, Cheryl, 6, 7, 156, 157n9
Bradford, Richard, 46, 55
Bristol Poetry, 600 Years of, 51
British Geological Survey, 25n16
Broackes, Justin, 11, 186n4, 188, 192, 206n26, 211n27
Broad, Charlie Dunbar, 187–189, 187n6, 188n7
  The Mind and its Place in Nature, 188
Brontë, Charlotte, 151n6
Brooke, Rupert, 107

'The Old Vicarage, Grantchester,' 107
Brophy, Brigid, 98, 98n3, 101, 101n5, 126
  Animals, Men and Morals, 101n5
  'In Pursuit of a Fantasy,' 101n5
  'The Rights of Animals,' 101n5
Browning, Gary, 30, 206
Bryson, Scott, 57, 72
Buddhist/Buddhism, 133, 164, 170, 175, 175n12, 226
Buell, Lawrence, 20n13, 161, 165, 166
Burnside, John, 175

C

Camus, Albert, 192
Capra, Fritjof, 228–230, 228n2
  The Tao of Physics, 228
Caprioglio Panizza, Silvia, 73n21, 130n14, 132, 199n21
Carson, Rachel, 8, 9, 162, 164
Chappell, Sophie Grace, 28, 194, 194–195n15
  Epiphanies: An Ethics of Experience, 194
Chatto & Windus, 53, 87
Cherwell newspaper, 52, 71n19, 74
Cherwell, River, 98
Chiasm, 197, 199
Christianity, 75, 134, 175
Clare, John, 5, 19, 47, 116
Clark, Timothy, 26, 27, 163, 189
  derangements of scale, 163
Climate
  breakdown, 58, 61
  crisis, 4, 5, 26, 58
  emergency, 5
Coelacanths, 159

Cohen, Margaret, 157, 158
Coleridge, Samuel Taylor, 46–48, 51, 70, 150
  'Dejection: An Ode,' 47
  'The Eolian Harp,' 48
  'Frost at Midnight,' 48
Comenetz, Michael, 147, 148n2, 149
Commoner, Barry, 58, 58n11
Conradi, Peter J., 43n1, 85, 86, 102–104, 114, 115, 159, 160, 200, 201, 201n23, 208, 225, 231, 234
Conservation, 175–178, 197
Conservative Party, The, 231
Cooper, David, 202
Coupe, Laurence, 160, 161, 165
Crabs, 159
Cronon, William, 209
Cue magazine, 54
Cynomorphism, 234

D

Day-Lewis, C., 52
Deep ecology, 18, 18n11, 56, 226–230
  deep ecologist, 227, 229, 230
Defoe, Daniel, 153
  Moll Flanders, 153
Denham, A. E., 11, 197, 198
Derrida, Jacques, 32, 228
Diamond, Cora, 3
Diaper, Jeremy, 68
Dickens, Charles, 6
Dipple, Elizabeth, 106, 107, 109, 148, 149, 156, 162
Disused railway, 214, 215, 216n28
Dogs, 67, 139, 199n21, 233, 234

Index

Dooley, Gillian, 15, 121n10, 204n25, 231, 232, 232n5, 232n6, 233n7

E

Earth, 2–4, 13, 18–21, 23, 25–28, 50, 58, 60–62, 64, 65, 69–71, 73, 74, 76, 78, 81, 83, 96, 96n2, 118, 127, 128, 130, 162, 183, 202–204, 213, 217, 219, 225–227, 236
Easterlin, Nancy, 154
Ecocentrism (biocentrism), 22
Ecocriticism, 4, 16–24, 34, 160, 161, 166, 185, 185n2, 236
  See also Environmental consciousness, environmental criticism; Material ecocriticism
Ecofeminism, 18
Ecological thought, 236
Ecology
  ecological consciousness, 3–5, 32, 33, 45, 95, 154, 225–237
  ecologism, 165, 166
Ecomarxism, 18
Econarratology, 154
Ecophenomenology
  ecophenomenological poetry, 56
  ecophenomenologist, 196, 200
Ecopoetics, 56, 88
  ecopoetry, 19, 56, 57
Ecosphere, 61, 72, 177
Ecosystem, 71, 77, 125, 161, 196, 211
Eliot, T. S., 67, 68, 85, 85n22

The Waste Land, 68, 85
Embodiment
  embodied affective response, 23, 194, 215
  embodied mind, 12, 183–219
  embodied self, 11
  embodied sensing, 189
Emotions, 12, 24, 45, 46, 111, 116, 125, 166, 166n11, 187, 190, 194, 198, 215, 234, 235
Empathy, 6, 24, 160, 190, 207, 235
Enactivism, 46n4
  enactive self-awareness, 194
Enlightenment, 138, 172
Environment, 6, 7, 11, 14, 17, 18, 20, 22–26, 28, 46n4, 49, 50, 56, 57, 70, 72, 74, 82, 123, 125, 126, 135, 140, 150, 152, 154, 159, 161, 162, 165, 167, 169, 172, 177, 184, 185, 188, 189, 194, 196, 197, 200n21, 203, 205, 207, 210, 212, 214, 218, 226, 228, 229, 237
Environmental consciousness
  environmental crisis, 16, 17, 233
  environmental criticism, 33, 35 (*see also* Ecocriticism; Material ecocriticism)
  environmental destruction, 232
  environmental discussions, 236
  environmental ethics, 31, 131, 132
  environmental imagination, 5, 13, 21, 24, 29, 55, 154, 226, 230, 237

environmental philosophy, 184, 230–233
environmental poetry, 56
Epiphany, 28, 88, 96, 97, 112, 122, 140, 194, 206–208, 226, 236
European Association for the Study of Literature, Culture, and the Environment (EASCLE), 16n10
Existentialism, 136n18, 191, 191n11, 192, 198
existentialist, 31, 151, 152n6, 168, 174, 191, 192

F

Fiddes, Paul, 74, 77, 201
First-person narrator, 149, 150
First World War, 68
Foot, Philippa (Pippa), 98, 176, 186
Fredriksson, Antony, 199n21, 233
Friedrich, Caspar David, 151, 151n4, 151n5, 151n6, 152, 157, 173
  Der Mönch am Meer, 157, 157n10
  Der Wanderer über dem Nebelmeer, 151, 151n4

G

Gaia hypothesis, 229
Garrard, Greg, 17, 18, 18n11, 56
Geological time, 26, 27, 71, 163
Georgic writing, 26
Ghosh, Amitav, 27
Gibson, Prudence, 211
  plant thinking, 211

Gifford, Terry, 56
Glotfelty, Cheryll, 16
Glover, Stephen, 29
Gollancz, Victor, 51
Good, the, 7, 12, 74, 76, 95, 106, 107, 124, 125, 155, 166n11, 175, 178, 201, 225, 226
Gothic, the, 116, 119, 121
Greene, Graham, 52
Green movement, the, 225
Greenpeace, 175
Green poetry, 56
Greenwich Theatre, 54
Gregg, Melissa, 184
Guardian newspaper, The, 10, 225
Gumbrecht, Hans Ulrich, 25
Gunnera manicata, 68

H

Haraway, Donna, 115
Harpers & Queen, 53
Hartley, L. P., 52, 170–173
Harvey, Graham, 113, 114
Haunts of the Black Masseur, 175
Head, Dominic, 19, 22
Heidegger, Martin, 96, 191, 197, 205, 206, 206n26
Heise, Ursula, 153, 155
Heptonstall, Geoffrey, 45
Hill, Susan, 53
  People: An Anthology, 53
Hinduism, 226
Holistic processes, 58
Hopkins, Gerard Manley, 5, 74–78, 81, 83, 84, 217
  'Binsey Poplars,' 81

Hopkins, Gerard Manley (*cont.*)
  curtal (sonnet), 76
  'God's Grandeur,' 78
  inscape, 77
  instress, 77, 78
  'Inversnaid,' 81
  journal, 75, 77
  'Justus quidem tu es Domine,' 217
  nature's laws, 78
  'The Nightingale,' 83
  'Pied Beauty,' 76
  'The Sea and the Skylark,' 81
  'Spring,' 74
  'The Windhover,' 74, 76
Horner, Avril, 98n3, 116, 124, 193, 198
Horticulture, 68
Hullah, Paul, 44–46, 45n2, 52, 53, 55, 66–69, 67n17, 74, 75, 87
Human exceptionalism
  human exploitation, 115
  human insignificance, 28, 29, 62, 164, 217
  human primacy, 21, 23, 96, 159, 191, 197, 236
Husserl, Edmund, 191, 197, 199, 199n20
Hutchings, Kevin, 47, 58
Hydrophasia, 158

I

Industrial revolution, 25
Influenza pandemic, 68
Interment, 67, 68
International Tree Foundation, 175
International Union of Geological Sciences (IUGS), 26
Intersubjectivity, 199n21, 233
Iovino, Serenella, 24, 135, 185
Ireland, 6, 15, 43n1, 97, 115, 154
Iris Murdoch Collections, the, 3, 3n4, 32, 44, 53, 53n9, 74, 75, 101n5, 147, 227n1
Iris Murdoch News Letter (*IMN*), 88
Iris Murdoch Review (*IMR*), 54, 62, 63, 87, 99, 100, 129n12, 202, 203, 209

J

James Tait Black Memorial prize, the, 54
James, Erin, 24, 153, 154, 158, 162, 164, 169, 170, 172–174, 203, 205
James, Henry, 103, 104, 202, 203
Jarrett-Kerr, Father Martin, 75
Johns-Putra, Adeline, 24, 27, 160, 162, 163
Jordan, Julia, 129, 130
Joyce, James, 5
Justice, 22, 95–97, 108, 126, 197, 231

K

Kant, Immanuel, 13, 47, 48, 121n8, 174, 199, 199n19, 199n20
Keats, John, 46, 85, 155n7, 174
Kermode, Frank, 27, 109
Kern, Robert, 154
Kerridge, Richard, 16, 17
Kestrel, 15, 63, 73–75, 79–81, 125, 131, 184, 225, 234
Kierkegaard, Søren, 192
Kinship, 164–169

Kuehl, Linda, 114

L
Labour Party, the, 231
La Cassagnère, Mathilde, 234
Ladino, Jennifer, 23
Language, failure of, 228
League of Nations, The, 19
Leeson, Miles, 32, 32n24
Les Temps Modernes, 190
Lidderdale, Hal, 45
Lidström, Susanna, 56
Light pollution, 84
Lipscomb, Benjamin, 151, 151–152n6, 191n11
Listener, The, 53
Logical positivism, 186, 198
London, 6, 7, 14, 20, 47, 85, 86, 99, 105, 106, 138, 158, 173, 174, 186, 211, 213
Lovibond, Sabina, 202, 203, 203n24, 209

M
Macfarlane, Robert, 63, 84, 151n5, 194–195n15, 217n28
MacLeish, Archibald, 2
Magee, Bryan, 29, 44
Mahon, Derek, 167, 168
  'The Seaside Cemetery (after Valéry),' 167
Marder, Michael, 211, 212
Marine environment, 161, 169
Marland, Pippa, 19, 22, 23
Martin, Priscilla, 51, 87, 201, 208, 209
Marvell, Andrew, 102

'The Garden,' 102
Marxism, 231
Material ecocriticism, 23, 34, 185
  See also Ecocriticism; Environmental consciousness, environmental criticism
McGrath, Hugh P., 147, 148n2, 149
McKibben, Bill, 177
Medlin, Brian, 230–232, 232n5, 232n6
  Human Nature, Human Survival, 230
Memory, 1, 59, 60, 68, 86, 100, 103
Men of the Trees, The, 175, 237
Merleau-Ponty, Maurice, 21, 23, 34, 136, 183–219
  corporeity of perception, 194
  ipseity, 196
  Phenomenology of Perception, 190–191
  Signs, 193, 193n14, 200, 200n22
  Visible and the Invisible, The, 193, 193n14, 196
  wild being, 34, 191, 197
Milligan, Tony, 124, 125, 234
Milton, John, 85, 85n22
Mind-body problem, 187
Moden, Rebecca, 137n19, 156n8
Moral agency
  moral perception, 4, 8, 9, 11, 13, 197
  moral philosophy, 7, 11, 30, 43, 57, 175, 187
  moral psychology, 188
Morel, Eric, 153
More-than-human, 21, 24, 61, 65, 71, 97, 113, 120, 131, 132, 135, 160, 166, 209, 210

Morris, Jackie, 217n28
Morton, Timothy, 20, 162, 163
  hyperobjects, 163
Motorway, 79–81, 86, 87
Mullan, John, 153
Murdoch, Iris, 1–35, 43–88, 95, 147, 183, 225–237
  and Alzheimer's disease, 98n4, 140, 232
  and art, 5, 7, 13, 14, 27, 29–31, 44, 46, 54, 56, 76, 79, 105, 107, 110, 137, 147, 151, 154, 156, 199, 201, 202, 207, 209, 228, 235
  and beauty, 2, 3, 9, 12, 13, 15, 74, 76, 77, 81, 85, 86, 97, 105, 107, 110, 125, 126, 128, 137, 156n8, 161, 170, 190, 201n23, 204n25, 215, 217n28, 219, 235
  featured poems
    'August,' 80
    'The City in the Plain,' 81–82
    'Edible Fungi,' 67n16, 68, 69, 69n18
    'Gunnera,' 66–68, 67n16
    'Rendezvous with Nightingales,' 84, 85
    untitled, 'Brilliant oaktrees, gold mixed with their green,' 64
    untitled 'for WR,' 'The trailing stars tell of dooms,' 62, 99
    untitled, 'Gently I have touched the thin lids of your eyes,' 59, 59n13
    untitled, 'The morning fills my eyes & my heart,' 61
    untitled, 'Over the wispy yellow slopes of the motorway,' 79, 80
    untitled, 'Snowdrops smell,' 66
    untitled, 'What does it matter,' 61
    untitled, 'You take life tiptoe. Too swift. Your splendid feet,' 70, 71n19
  further poems
    'The Brown Horse,' 67n17
    'The Coming of April,' 51
    'Conversations with a Prince,' 55, 87
    'Lower than the Angels,' 51
    'The Phoenix Hearted,' 46, 51
    'Poem,' 52, 52n8, 74
    'Star-Fisher,' 51
  and the Good, 7, 12, 30, 74, 76, 106, 107, 178, 201, 225, 226
  and the journals, 3, 3n4, 3n5, 32, 32n24, 44, 53, 54, 74, 75, 78, 97, 98, 132, 185, 186, 188–190, 194, 236, 237
  and the kestrel, 15, 63, 73–75, 79–81, 125, 131, 184, 225, 234
  letters, 32, 45, 97, 98, 101, 101n5, 139, 188, 190, 229, 230, 232, 237
  novels
    The Black Prince (*BP*), 6, 54, 231
    A Fairly Honourable Defeat (*FHD*), 216n28, 217n28

The Flight from the Enchanter (*FE*), 33, 95–141
The Good Apprentice (*GA*), 6, 10, 34, 60, 104, 185, 185n3, 200, 203–219, 233
The Green Knight (*GK*), 6, 14, 15, 34, 127–129, 234
Henry and Cato (*HC*), 79
'Jerusalem,' 53
The Nice and the Good (*NG*), 60, 100, 127, 154, 193–196, 198, 206
Nuns and Soldiers (*NS*), 4n6, 6, 33, 60, 95–141, 154, 163, 210
The Philosopher's Pupil (*PP*), 6, 133, 229, 229n4, 230, 234
The Sacred and Profane Love Machine (*SPLM*), 86, 87
The Sandcastle (*TS*), 9, 10
The Sea, The Sea (*TSTS*), 6, 11, 34, 54, 65, 100, 101, 126, 127, 147–178, 156n8, 183, 234
The Unicorn (*TU*), 6, 33, 95–141, 147, 201, 210
An Unofficial Rose (*UR*), 33, 95–141
other writing
  'Millionaires and Megaliths,' 129n12
  The Murdochian Mind, 31, 185n3, 193, 228, 234
  'Taking the Plunge,' 176n15, 176n16
philosophy
  'Against Dryness' (AD), 47, 120, 136n18
  'The Existentialist Hero' (EH), 193

'The Fire and the Sun' (FS), 30, 110
Gifford Lectures, The (Giffords), 1, 198, 199n20
'The Idea of Perfection' (IP), 8, 24n15, 110
'Literature and Philosophy: A Conversation with Bryan Magee' (LP), 29n22, 30, 44
Metaphysics as a Guide to Morals (*MGM*), 1, 1n1, 2, 2n2, 13, 14, 32, 95, 99, 100, 125, 136, 161, 174, 198–200, 199n20, 228, 229, 235, 236
'The Novelist as Metaphysician'(NM), 152n6, 186, 191, 191n11, 192, 198
'On "God" and "Good"' (OGG), 7, 7n8, 12, 57, 97, 124, 211n27
Sartre: Romantic Rationalist (*SRR*), 31, 192, 198n18
The Sovereignty of Good, 4n6, 8
'The Sovereignty of Good over Other Concepts' (SGC), 15, 47, 50, 63, 73, 75, 76, 79, 81, 121n8, 125, 126, 130, 131, 161, 211n27
'The Sublime and the Beautiful Revisited'(SBR), 5, 120, 120n7, 121, 121n9
'Thinking and Language'(TL), 12, 13, 55, 105
play
  'The Servants and The Snow,' 54

Murdoch, Iris (*cont.*)
  and the poetry, 4, 5, 9, 19, 20, 27, 32, 33, 43–46, 50, 52–56, 58, 63, 75, 76, 79, 87, 88, 132, 141, 147, 166n11, 226, 236, 237
  poetry notebooks, 3, 3n4, 32, 44, 45, 52, 53, 53n9, 59, 75, 81, 237
  and swimming, 139, 159, 167, 175, 176, 176n16
  and unselfing, the unself, 15, 28, 57, 96, 130, 130n14, 131, 161, 234
  worldview, 18, 21, 22, 225
Muroya, Yozo, 44, 66, 75, 87

N

Naess, Arne, 227
Natural world
  nature, 4, 44, 96, 151, 185, 225
  nature poetry, 56–58, 75, 88
Neuropsychologists, 235
Neuroscience, 189, 190
New College, Oxford, 232
Newnham College, Cambridge, 186, 187n6
The New York Review of Books, 175, 176n15
Nicol, Bran, 117, 124
Nietzsche, Friedrich, 174
Nixon, Rob, 23, 175n13
Noise pollution, 84
Nonhuman/non-human, 17, 19–21, 23, 27, 34, 56, 57, 82, 113, 115, 122, 124, 125, 129, 130, 135, 150, 158, 161, 165, 166, 185, 199n21, 208, 233–235
  agencies, 115, 122
  animals, 233–235
  creatures, 57, 158
  interaction, 135
  life, 124
  representation, 161
  subjects, 82
  world, 19, 124, 130, 150
Nuclear
  power plant, 10
  war, 9, 10
  weapons, 10, 228
Nussbaum, Martha, 24, 24n15

O

Oak Taylor, Jesse, 24, 25
O'Flinn, Paul, 58
Oliver, Tom, 71, 72
Oppermann, Serpil, 24, 135, 185
Otherness (of nature), 73, 74
Other, the, 7, 199, 199n21
Oulton, Lucy, 4n6, 12, 21, 31, 185n3, 189, 193, 228, 228n2, 229n4
Ovid
  'Leda and the Swan,' 14
  'Tereus and Philomela,' 85n22
Oxford Forward, 52

P

Parham, John, 77
Phenomenology, 11, 34, 190–193, 197, 199, 199n21
  phenomenologists, 31, 190, 191, 197, 198, 211

Place, ethics of, 4, 6, 24, 50
Plato, 211, 211n27
　Timaeus, 211
Pliatzky, Leo, 192, 192n13
Plumwood, Val, 124, 176
Podcasts, 1n1, 104, 109, 176n16, 237
Poetry
　ecophenomenological poetry, 56
　ecopoetry, 19, 56, 57
　environmental poetry, 56
　green poetry, 56
　nature poetry, 56–58, 75, 88
Poetry London/Apple Magazine, 66
Poet Venturers, 51, 52
Powers, Richard, 27n21
　The Overstory, 27n21

Q

Queneau, Raymond, 43, 98, 187n6, 190, 190n9, 191

R

Raban, Jonathan, 157
Racine, Jean, 130, 131n15
　Phèdre, 130
Rainbow Warrior, The, 175
Read, Daniel, 30
Realism, 6, 7, 14–16, 24, 27, 28, 103, 120, 160, 184, 200, 202, 203, 236
Regeneration, 51, 66, 68–71, 82, 83
Reilly, Evelyn, 56
Remorse, 170, 172, 173, 213, 218
Robson, Wallace, 62, 63, 87
Romanticism, 46–48, 140, 151, 152, 157, 166, 174
Rosa canina, 104, 110, 112

Rose, W. K., 104, 200
Rossetti, Christina, 67, 67n17
　'Spring,' 67, 74
Rowe, Anne, 6, 7, 13, 14, 21, 51, 54, 55, 87, 98n3, 104–106, 109, 128, 156, 160, 168, 193, 198, 201, 205, 208, 209, 237
Rückenfigur, 151, 151n4, 157
Rueckert, William, 16
Ruskin, John, 49, 166

S

St. Anne's College, Oxford, 186
Sale, Roger, 103
Salwa, Mateusz, 103, 211, 218
Sargasso Sea, The, 176
Sartre, Jean-Paul, 31, 45, 174, 190–193, 191n10, 192n12, 198, 205
　La Nausée/Nausea, 192
Scale
　scalar dimension, 27n21, 161
　space and time, 33, 97
　temporal layers, 163
Schopenhauer, Arthur, 174
Scientism, 8, 57, 227
　See also Anti-scientism
Seals, 157, 157n9, 159, 234
Sea monster, 158, 168
Seascape
　seas, 158, 176
　waves, 18, 20n13, 22, 77, 138, 163, 164, 167
Second World War, The, 58
Seigworth, Gregory, 184–185
Sekula, Allan, 158
　hydrophasia, 158

Sellafield, 9, 10
Sensa, 187, 188
Sensation
   sense perception, 21, 197, 202, 218
   sensing human body, 114
   sensory body, 189, 190, 197, 201
   sensory experience, 4, 13, 100, 189
Shakespeare, William, 8
   As You Like It, 124
   Duke Senior, 124
Shelley, Percy Bysshe, 46
Shepherd, Nan, 195n15
Silesius, Angelus, 63
   'Die Rose ist ohne Warum,' 63
Slow violence, 23
Smallwood, Norah, 53
Smith, Mick, 130–132, 134
Snow, C. P., 8, 8n9
Solipsism, 139, 178, 195, 235
Somatic
   history, 188
   response, 194, 197
   sense history, 188
Somerville College, Oxford, 81
Soper, Kate, 20
Species, 3, 11, 23, 28, 101n5, 105, 165, 227
Spender, Natasha, 133–135, 134n17, 137
Spender, Stephen, 133
Stephen, Leslie, 50
Stimmung, 25
Stone, Reynolds, 79, 128
Stones
   lithic materiality, 97, 140
   rock, 24, 125–128, 135–137, 139, 140, 158, 159, 163, 165–167, 173, 177
Storyworld, 22–28, 154
Stratigraphy, 25n17
Sublime
   Romantic sublime, 115, 158, 195
   sublimity, 112, 121, 134, 151n6, 235
Supernatural, 122, 164, 202, 235
Swimming, 139
   See also Murdoch, Iris

T

Taoism, 226, 228–230
Technology, 2, 9, 10, 57, 227, 228
Television, 2, 2n2, 10, 98n4, 109
Thales of Miletus, 162
Thames, River, 14, 34, 82, 97, 99, 176
Thatcher, Margaret, 231
Tintoretto, 104–106, 111
   Susannah Bathing, 105, 106
Todd, Richard, 202
Trees, 3, 4, 23, 24, 49, 50, 57, 72, 81, 82, 98, 98n4, 121, 122, 124–126, 134, 136, 138, 159, 184, 194, 203, 205, 209–212, 216, 217
Tuism, 195, 195n16
Tylor, Edward, 113

U

United Nations Relief and Rehabilitation Association (UNRRA), 43n1, 45

Urban encroachment, 84, 86
  of soundscape, 84

V

Valéry, Paul, 147–149, 152, 167, 168, 170
  'Le cimetière marin,' 148, 149, 162, 167
Vegetal, 34, 71, 73, 95, 97, 140, 210, 211, 214–216, 218, 219
  agency, 34, 183–219
Vidal, John, 10
Violence, 23, 102, 119, 168–172

W

Walton, Samantha, 72, 72n20, 73
  'four apocalypse sonnets,' 72
Water, 14, 15, 34, 49, 69, 70, 72, 82, 99, 124, 136–140, 138n20, 148, 150, 155–160, 155n7, 157n9, 162, 167, 172, 175–178, 212–215, 233
Waugh, Evelyn, 52
Waugh, Patricia, 11, 12, 189, 190
Weather event, 115, 117
Weik von Mossner, Alexa, 24, 185, 185n2
Weil, Simone, 15, 130, 130n14, 131, 131n15, 174
  décreation, 15, 130, 130n14

Weinberger, Harry, 130n13, 137n19, 156n8
Westling, Louise, 23, 196, 197
White, Frances, 4n6, 15, 32n24, 62, 63, 87, 104, 114, 218, 236
White, Gilbert, 65, 159, 213
White, Lynn, 134, 135, 137
Wild being, 34, 191, 197
  See also Merleau-Ponty, Maurice
Williams, Raymond, 18, 19, 24
Wilson, Angus, 103
Wittgenstein, Ludwig, 101n5, 126
Woods, Derek, 26
Woolf, Virginia, 3n5, 5, 6
Wordsworth, William, 5, 19, 46, 49–51, 70, 87, 98, 140, 166, 166n11
  Lyrical Ballads, 46
  'The Wanderer,' 49, 50
Wytham Wood, 98n4

Y

Yeats, W. B., 5, 15, 51, 56
  'Easter 1916,' 15
  'Lapis Lazuli,' 51
  'Leda and the Swan,' 14
Yilmaz Kurt, Zeynep, 228

Z

Zoofocalization, 234
Zoopoetic, 234

Printed in the United States
by Baker & Taylor Publisher Services